KU-324-721

INTO THE TANGLED BANK

In which our author ventures outdoors to consider the British in nature

Lev Parikian

Elliott&Thompson

First published 2020 by
Elliott and Thompson Limited
2 John Street
London WC1N 2ES
www.eandtbooks.com

ISBN: 978-1-78396-506-9

9 8 7 6 5 4 3 2 1

A catalogue record for this book is available from the British Library.

Typesetting by Marie Doherty
Printed in the UK by TJ International Ltd

To my parents.

CONTENTS

It is interesting to contemplate a tangled bank, clothed with many plants of many kinds, with birds singing on the bushes, with various insects flitting about, and with worms crawling through the damp earth, and to reflect that these elaborately constructed forms, so different from each other, and dependent upon each other in so complex a manner, have all been produced by laws acting around us.

Charles Darwin, *On the Origin of Species*, 1859

INTRODUCTION

I'm lying, as you do, on the pavement.

It's comfortable enough, as pavements go. I've managed, for once, to get myself into a position – lying on my right side, propped up on one elbow – that doesn't have parts of my body screaming in protest.

This isn't a random act. I'm not given, when walking down the street, to plonking myself down willy-nilly with a cheery 'Time for a nap!', regardless of the needs of my fellow pedestrians. I do have a reason.

The pavement in question is on the street corner opposite our house in South London. It's the kind of corner you might find in any town or city. It's deeper than many pavements because of the postbox, and the Victorian stone drinking fountain set back a few feet from the road. But otherwise it's normal enough. The road is main-ish, and when the traffic lights at the top turn green, there's a steady stream of cars and bikes and buses trundling down the hill, but sometimes, just for a minute, all is peaceful and you can hear the soft throatiness of a wood pigeon or the merry tinkling of a goldfinch above the background London hum.

Behind is a large vacant lot, ripe for development but somehow neglected, closed off with an untidy wooden hoarding that leans drunkenly inwards – much complained about by the locals, occasionally daubed with graffiti. Every so often someone dumps an old sofa or a wonky IKEA bookcase or a broken TV here. It's that kind of corner. Despite the reassuring presence of a few mature trees, it's a nondescript place – plain, unwelcoming, not an invitation to linger.

Until earlier this year.

Earlier this year the council hatched a scheme to spruce it up a bit, the kind of scheme – increasingly common – that gives one a shred of hope for humankind. A group of people had had enough of squalor, and pulled together to create a little nature oasis, something to brighten up the neighbourhood.

This corner is a special place for us, not just because it's local, but because my wife designed it. Some well-placed planters, a few raised flagstones, all imaginatively planted – instant regeneration on a small scale. The kind of thing that makes a difference. Now, when you pop out to post a letter, you can say hello to *Verbena bonariensis*, *Stipa gigantea* and *Linaria purpurea* rather than a scattering of fried-chicken cartons, empty cans of Stella, and an inexplicable shoe.

On this occasion I have, as it happens, popped out to post a letter, and as always I take a moment to check on the plants, to see how they're doing. Someone has used the low planter by the fountain as a bench again, gently squashing the ivy. Perhaps it's time for a handcrafted sign along the lines of one I saw a couple of weeks earlier in another part of town, a piece of A4 in a plastic sleeve, lovingly decorated in large, childish lettering: 'PLEAS do not sit here it is NOT A BENSH'.

I cross the road, and my eye is caught by a movement. A fluttering, scattery movement, as of one not quite in control but going there anyway. As of, in fact, a butterfly.

I trace its skeetering path, and watch as it comes to rest on one of the plants. A year ago, it would have flown on, might not even have been lured to this corner. Today it's stopped for a rest and a drink. It settles on a flower, looking set for the medium haul. A motionless butterfly is always worth closer examination. In flight they're eye-catching but often frustratingly transient, flitting away on apparently random zigging paths, and usually out of my sight within a couple of seconds. But when they settle, it's nice to move in for a closer look.

And that's when I lie down on the pavement. Because, you know, why not?

I manage to do so without casting my shadow over it and making it fly away. So far so good. I inch closer – and to hell with my trousers, they can always be washed – until I'm no more than two feet from it. It stays put. I feel trust has been established.

It is, to my eyes, a thing of transcendent and eye-catching beauty. In flight, a delicate wisp of ultramarine, insubstantial-looking, vulnerable to the merest puff of wind; and then, settling on a purple flower whose name I don't know, it's transformed, a delicate patchwork of white and black spots on a soft beige background with orange teardrops at the wing edge, the only hint of blue a silvery tinge at the base of the underwing. Fine, intricate, fascinating.

The kind of thing, in fact, you might lie down on a pavement for.

I don't know about butterflies.

That's a quarter of a lie. I didn't know anything at all about butterflies (cabbage white, umm, that's it) until the resurgence of my interest in birdwatching a few years ago. A scant grasp of butterfly identification came along with it, almost as collateral – birds go quiet in July and August, just when butterflies are at their peak, so it's a natural progression. Now I can recognise a few species almost unfailingly (brimstone, orange tip, comma, peacock, speckled wood, meadow brown), and some with a bit of help from a field guide, as long as they sit nice and still, which they don't. But whole swathes of them remain mysterious, especially those orange and black masters of disguise, the fritillary family, damn their fluttery wings.

This one, though, I know. It's a common blue. Or, so my field guide tells me, *Polyommatus icarus*. The male of the species. Widespread in the British Isles, in flight from May to September.

That word 'common' doesn't do it justice. It dismisses it as something nondescript, unimportant, not really worthy of attention. But here it is in front of me, transfixing me with its delicate charms.

I become aware of a movement, and look up. A small girl, four or five, running up the hill towards me. Behind her, a man – presumably her father – making slower progress. Phone out, head down, obviously trusting to peripheral vision that everything's fine and that his daughter's not going to die a grisly death under the wheels of a truck.

The girl stops a few feet away from me, perplexed. And now the man looks up from the screen, takes in the situation, and calls across, at the same time veering away from me to turn the corner. He beckons to her.

'Come on, poppet.'

I can't blame him. A middle-aged man lying on the pavement is very much the kind of thing you might want to keep your daughter away from.

But it seems a shame. What I really want to do is call over, summon both of them across so they can share this everyday miracle, the casual and phenomenal beauty of a common thing, so easily overlooked.

'Look at this. Isn't it beautiful?'

It overtakes me sometimes, this urge to share. There's nobody quite so zealous as the recent convert, after all. At times I find myself fired with an almost embarrassing missionary zeal, enthused by a mundane and passing spectacle. I see a feral pigeon execute a particularly impressive landing, or a silver birch with finely coiled peeling bark, or a dragonfly just being a dragonfly, and part of me wants to yell, 'Look at this, everyone. LOOK AT THIS!'

But no matter how enthralled I am by the butterfly, I am also aware of how strange I must look. So I stay silent, restrained by a self-consciousness of which I'm slightly ashamed, and off they go, the girl, pleasingly old-fashioned, skipping along the pavement, the man, inevitably modern, buried once more in his phone. After a few seconds the butterfly flits off anyway, and I go home for a coffee, and it's only an hour later that I realise I forgot to post the letter.

They stay in my head. The girl, the man, the butterfly. Something about the juxtaposition sticks: my extreme enthusiasm, the girl's curiosity, phone guy's obliviousness. Three levels of engagement.

I remember the moment with mild regret. Imagine I had called them over. What was the worst that could have happened? He might have ignored me completely; he might have come across, looked at the butterfly with polite interest, nodded a bit, said, 'Oh yes', then made an excuse and left, forgetting about the incident almost immediately. But maybe, just maybe, the little girl might have remembered it, noticed the next butterfly, and the next one, and an interest would have been seeded.

And in any case, it would have been nice to show it to them. As they were there and everything. It really was startlingly beautiful.

Not that I would have had anything profound to say beyond 'Look at this gorgeous butterfly.' As I say, I don't know about butterflies. I am, truth be told, a nature-watching Johnny-come-lately. Ten years ago I was like phone guy, head down, looking at the natural world, if at all, through unseeing eyes. But now I'm the man in the park staring through binoculars, examining the lichen on the tree trunk, pointing at the swifts as they re-enact *Top Gun* above my head.

Something in me shifted in that time, something fundamental. Decades had passed between my birdwatching-obsessed childhood and the resurgence of interest in my late forties – fallow decades during which the natural world more or less passed me by.

It's not quite true to say I was blind to it all. I didn't walk around with a cardboard box on my head, oblivious to the natural world around me. But nor did it engage my attention in anything but a cursory way. Yes, I enjoyed going for walks in the countryside – particularly if there was a pub or restaurant at the end – and yes, I was aware of wildlife if it thrust itself upon me: rutting deer in Richmond Park; an urban fox, late at night, trotting along the road as if it owned it; the wren that flew in through the kitchen window that time. But I would no more have gone to a wildlife reserve than I would have danced a merengue wearing an Eeyore onesie while eating a tub of desiccated coconut.* It simply didn't occur to me that this would be a worthwhile thing to do.

But the interest, seeded during a childhood in a small village in Oxfordshire, was merely dormant. And sometimes things sneak up on you. As middle age exerted its flabby grip, I found myself, almost without realising it, noticing birds again. And not just noticing them but reading about them and relishing them and going out of my way to find them. And before I knew it, they possessed me utterly, consuming more time and energy than many people would consider necessary.

I wrote a book about it, a description of one year in the grip of this obsession. Some people, I think, assumed it was a passing thing,

* I do not like desiccated coconut.

a convenient peg for a humorous book about being a middle-aged man, and that having written it I would return to some semblance of normality, whatever that means. But no. It was just the beginning.

If I describe myself as a birder, then that's because birds have been the primary focus of my nature-watching activities. But they can't exist in isolation. It struck me soon enough that there was a certain irony in the disconnect between my depthy knowledge of a bird – its scientific name, size, plumage details, distribution, favoured habitats, migration patterns, nesting habits, diet, clutch size, and even its name in Finnish – and the graphene-thin extent of my knowledge of the tree it was perched on. And almost everything else, for that matter. It was all about the birds. The other stuff – the trees, flowers, butterflies, lichens, otters, moths, foxes, fungi, hares, lizards, frogs, grasses, dolphins, bats, weasels, arachnids, fish, bees, beetles, seals, dragonflies, mosses, voles and ladybirds – well, they were lovely, of course, but they weren't birds, were they? I couldn't get obsessed by them.

Or could I?

It turns out I could.

One thing leads to another, and now, almost without realising it, I've accumulated a selection of field guides. Guides to wildlife in general, guides to UK butterflies* and moths,† guides to trees, wild flowers, insects and much more; where to see them, how to watch them, what they do, how and when and why they do it.

Going out to be in nature is now an unbreakable habit, a way to nourish body, mind and soul all at once. And nearly everywhere I go there are other people, each experiencing nature in their own

* 60 species or so.

† 2,500 of them YOU ARE KIDDING ME HOW DO PEOPLE MANAGE?

way. Birders, ramblers, joggers, cyclists; dog-walkers, gardeners, photographers; loving couples, grumpy singles, nature groups, families; kite-flyers, frisbee-flingers, den-builders, grass-loungers; young whippersnappers, old farts, middle-aged ne'er-do-wells; millennials, boomers, generation X-ers, generation Y-ers, generation somewhere-in-betweeners; experts, beginners, specialists, all-rounders, or just people out for a stroll in the sun.

A lot of them don't realise they're experiencing nature, of course. They don't tell their partners, 'Just off out to experience nature, darling. Back in ten.' No, they've just popped out to the corner shop or are on their way to work or meeting Vanessa for lunch or taking the dog out for a walk or just sitting in the sun with a pint and a packet of crisps. But whether oblivious or fully engaged, they're experiencing it all right. Because – and this is a staggeringly obvious point, but I'm going to make it anyway – nature is all around us. You can't get away from the damn stuff. And we are part of it, whether we realise it or not.

And yet we are, so we are constantly told, more remote from it than ever before. Our spreading conurbations gobble up what countryside is left, and that countryside grows ever more denuded of wildlife. We raise our children in concrete jungles, and they wouldn't recognise a cow if it sat on them, don't know of the existence of kingfishers or bluebells or acorns.

How do we square this, then, with the enduring and growing popularity of nature programmes such as *Planet Earth* and *Springwatch*? Are we merely armchair nature-lovers? Or are these programmes engendering an abiding interest in the natural world, inspiring us to go out and commune with it more intimately?

The truth, as always, is somewhere in between. Interest in nature is a spectrum – from oblivion to obsession and myriad stops on the way – and each person occupies their own particular place on it.

These people – what they do, where they do it, how and when and why they do it – have come to be an integral part of my own experience. And I find them, inveterate people-watcher that I am, beyond captivating. I know how *I* respond to nature, and I know how *I* enjoy relating to it. But look at that man there, the one standing next to the 'DO NOT FEED BREAD TO THE DUCKS' sign, feeding bread to the ducks. Fascinating. And that woman, standing oblivious at the zoo while her child communes with a penguin. Or, at the other end of the spectrum, the warden of a bird observatory, living off-grid on an isolated island for nine months at a time, recording everything, no matter what the weather. Who are these people? What are their quirks, their foibles, their own particular ways of responding to and interacting with the natural world around them? I don't want to climb either soapbox or high horse* about the parlous state of nature, but these questions feel more relevant than ever. Because when we lose contact with something, we stop caring about it; and when we stop caring, it's gone before we know what we've lost.

If normal people and their relationships with nature are spell-binding, then so too are the experts. It's one thing to notice a tree; another to be able to identify it from the fissures in its bark or the shape of its leaves; yet another to understand its function in the local ecosystem; and quite another thing entirely to devote your life to them, almost to the exclusion of everything else.

I find myself delving into the lives of the outliers: people who, at one time or another, have seen the world through different eyes,

* A really neat piece of contortion if you can pull it off.

whose exceptional observations and deeds, to this admiring amateur at least, are unimaginable in their scope and reach. I'm infected with a version of the same insatiable curiosity that led Gilbert White to explore the wildlife of his local patch, or Anna Atkins to press seaweed between glass and make the first cyanotypes, or Peter Scott to examine the individual facial patterns on Bewick's swans. Perhaps it's making up for lost time; perhaps it's a rebuke to my teenage self, that feckless do-nothing who rejected learning; perhaps it's because middle-aged me feels time is running out, is desperate to catch up. Whatever the reasons, I need to find out more about all of this: nature itself, our relationship with it, and the people who have contributed so much to our understanding of it.

This threefold motivation leads me to plan a journey, starting at home in London and from there meandering across Britain. It takes in the places closest to us – our houses and gardens and local patches – before striking out, to zoos, wildlife reserves, country parks and wildernesses. The journey's course is partly driven by links to the past, visiting the homes and neighbourhoods of the great and the good, whose achievements and discoveries lead me to examine what I see more closely and carefully. But I also want to study – however unscientifically – the people I meet on the way: their habits, their preferences, their ways of being in nature. And, of course, at the heart of it all is nature itself, in what Gilbert White called its 'rude magnificence'.

Tempting though it is to strike out directly to the farthest reaches of the British Isles in search of eagles, otters and dolphins, there are rich pickings to be had under my own roof. Less glamorous pickings, but pickings nonetheless. So this journey starts in the home, where – like it or not – there's enough nature to last you a lifetime.

1

HOME, SWEET HOME

In which our author goes toe to antenna with creepies, crawlies, jumpies and munchies – our nearest neighbours, for better or for worse

It's just sitting there, sizing me up, poised to pounce should the need arise. In normal circumstances the need wouldn't arise, and we could both get on with our respective days: me answering emails and going to the shops, and it sitting in the sink for a while before mysteriously disappearing and then coming out at night and dancing a jig on my slumbering face.*

But I need to do the washing up, so disturb it I must.

Some people, I dare say, would flush it down the sink; others – harder-hearted, crueller, but also quicker – might dispatch it with

* They definitely don't do this. OR DO THEY?

a quick blow from a suitable implement; my preferred method, honed throughout a childhood in a house with innumerable shady crevices, is to coax it onto a square of kitchen paper, and hope it doesn't make a bid for freedom up my sleeve while I carry it to pastures new.

Despite my assertion above, it isn't poised to pounce. Nor is it sizing me up. The giant house spider (*Eratigena atrica*) has eight eyes, for sure, but its grasp of detail is weak – if it's aware of me it's as an amorphous shape in the distance rather than as a looming existential threat. And while the set of its legs gives the impression it's ready to spring into action at the slightest provocation, its instinct is to retreat from humans, not launch itself into a frenzied attack. If its presence in our sink is motivated by anything other than lust, the likeliest reason I can think of is that it enjoys the cool feeling of porcelain on its feet.

This spider is most likely to be a male on the pull, roaming the highways and byways of our house on the lookout for a mate: 'Male *Eratigena atrica* seeks female same species. Must have: GSOH, willingness to mate repeatedly, innate desire to devour body of deceased partner. No time-wasters.'

Left to its own devices it'll leave the sink soon enough and find a nice neglected corner, where it'll spin a sheet-like web and wait for insects, quietly and undramatically playing its own small part in the management of our domestic ecosystem.

Keen to procrastinate, and even keener to make use of the nifty 'close-focus' binoculars I've recently bought for just such a purpose, I bend down and look at the spider more closely. There's a fascination in its absolute stillness, the asymmetric angles of its splayed legs, and the knowledge that those delicate limbs are capable of propelling it at a speed to put the fright up any watching humans.* Through the binoculars I can examine the subtle variations in its colouring – it's mostly dark brown, but the ends of its legs morph to black, and there are pale spots on its abdomen. Both these features are also covered with fine hairs.

It's a toss-up whether this close examination makes the spider more or less intimidating. Seeing it magnified to this extent brings to mind the old question of whether you'd rather fight one horse-sized duck or a hundred duck-sized horses, and the thought of substituting spiders into that old meme is enough to send shivers down even this arachnophile's spine. But I shove such thoughts to one side and concentrate merely on observing it.

The art of observation is elusive. Here is a thing. What does it look like? Sound like? Smell like?† What, if anything, is it doing? Why is it doing or not doing it? What might it do or not do next?

It's all too easy to be sucked into the superficiality of 'tick and move on', as if merely having seen something is in itself an achievement, somehow making you a better person. But for all that a sighting brings its own frisson, there's a deeper satisfaction to be found in a more detailed observation, even if the object in question, like this spider, is to all intents and purposes doing nothing.

* Half a metre a second, give or take, which scales up to a gazillion miles an hour. But these bursts of speed, one of the contributing factors to some people's fear of spiders, are understandably brief.

† Nature stimulates all the senses. This is why I refer to 'birding' rather than 'birdwatching' – do all those chirrits and wirbles and twiddly-doo-wops I've heard from the depth of the bush, without ever laying eyes on the bird, not count?

There's a lot to be said for the mesmerising quality of an animal – from heron to hedgehog – just being still, following the timeless zen advice: don't just do something, stand there. At the very least, the mere act of spending time in the creature's company lends familiarity; and this kind of familiarity, rather than breeding contempt, increases my level of comfort with this spider's presence, reduces its otherness, and makes me more prepared to accept it as part of my daily life. This thing before me is normal, not strange.

What is strange, or at least might appear so to the casual observer, is the sight of a middle-aged man staring at a pile of washing-up through a pair of binoculars, but in my absorption I'm comfortable with that thought. Let the record show my strangeness for all to see.

I put the binoculars down, fetch the kitchen roll and slide a single sheet deftly under the spider. It scuttles away for a second, but I persist, and soon it is sitting calmly enough on its new papery resting spot. Master anthropomorphist that I am, I toy briefly with giving it a name. Sid, perhaps. But that way madness lies. Start naming all the insects in the house and I'll run out of names before sundown.

I put it to one side and do the washing-up.

Whether we like it or not, we share our homes with other species: dust mites, lurking in our mattresses; silverfish, feeding off food scraps and scuttling off behind the skirting board; fruit flies, appearing as if from nowhere at the first sign of a half-mushy banana; daddy-long-legs, flapping about melodramatically near light bulbs; house flies, buzzing around aimlessly and failing to go out through the window you've pointedly opened for just such a purpose; ladybirds, yellowjackets, black ants, bedbugs, earwigs, centipedes, mosquitoes, weevils, cockroaches, woodworm, death-watch beetles, biscuit beetles, bacon beetles, cellar beetles, grain beetles, flour beetles, absolutely bloody everything beetles. Millions and millions of the little buggers, and that's without even mentioning parasites such as head lice* and fleas, or the eight-legged mites that make their homes and graves in our eyebrows, or the host of bacteria in our gut.

And breathe.

Whatever the species – and rest assured it's deeply unlikely you'll be infested with more than, ooh, let's say half a dozen of the above-named at any one time – all the evidence points to one conclusion: we're not that keen.

The reasons for this aversion are deep-seated. Ever since humans shelved their nomadic ways and settled into the routine of tilling the land and generally staying in the same place, we've created environments – dry, warm, food-filled, or a mixture of all three – that have appealed in some way to other species. And historically a lot of these species have been a nuisance: they've eaten our food, given us diseases and even destroyed our buildings. Some of them bite us, causing symptoms ranging from mild discomfort to death. Others wriggle or crawl or squirm or just sit there in our sinks, their very existence causing us discomfort or revulsion in ways we find difficult to articulate beyond saying, 'I just can't stand them', and giving a horrified shudder.

The extent to which we're averse to our uninvited guests depends partly on personal preference. We each have our own level of tolerance for ickiness, and apply it in different ways. Someone who quakes at the thought of an earwig might show surprising calmness when

* It turns out they *do* prefer clean hair – it's not a myth perpetuated by children with head lice to fend off mockery and ostracisation.

confronted with a bumblebee trapped in a window, for example; and a person whose reaction to a wasp's nest in the loft is the entirely rational 'well, they've been there all summer without causing us any problems, and they'll be gone in a couple of weeks, so I might as well leave them', might also be the first to start back in shock and revulsion at the sight of a mouse scurrying across the kitchen floor. The definition of 'pest' is different for everyone.

Quantity plays its part in shaping our reaction. Calm as I was about Sid's presence in the sink, my *sang* would have been somewhat less *froid* had he brought a couple of dozen friends along. And while I'm perfectly content to share my space with these octo-podded little scamperers – just as long as rapid scuttling is kept to a minimum* – I wouldn't go so far as to keep a spider as a pet; a position shared, I might add, by more than 99 per cent of the population. True arachnophobia – irrational and crippling fear, a condition not to be dismissed lightly – is rare, but spiders do cause anxiety in many people who, in their rational moments, will happily concede they know them to be harmless.† This anxiety is no doubt rooted in the remnants of some primeval instinct that prompts us, on seeing the dreaded foe, to RUN STEVE RUN FAR RUN FAST IT'S GONNA KILL YOU.

A similar level of distrust is attached to the house spider's cousin, the daddy-long-legs spider (*Pholcus phalangioides*).‡ Most commonly

* A rule, I should add, I apply equally rigorously to myself.

† In the UK, at least. There are many parts of the world where venomous spiders are common enough to be considered a threat to public health, but Britain isn't one of them. Even in Australia, embedded in British imaginations as a hotbed of fearsome arachnids, actual fatalities from spider bites are vanishingly rare.

‡ Not to be confused with the daddy-long-legs mentioned above, also known as craneflies.

seen sitting upside-down in the higher reaches of your kitchen or bathroom, these spiders appear from a distance to consist of legs, and legs alone. Their web – no more than a straggling pile of silky strands – looks shambolic, apparently having been thrown together at ten to five on a Friday afternoon by a disaffected designer who couldn't be bothered to read the brief. They might not crouch in the same menacing way as their cousins, as if coiled and ready to strike, but we're still prepared to believe the worst of them, which is a shame, because the daddy-long-legs spider is entirely harmless, and (like the house spider) helps control the population of flies – flies that we then, using that magnificently selective logic unique to humans, complain about.

This demonisation of spiders isn't helped by the newspapers, for whom all arachnids are bundled into a sinister bag marked 'hysteria trigger'.

The unfounded frothing about the various *Steatoda* species lumped with the name 'false widow spider' is a case in point. They are, it is true, venomous. But then so are all spiders, to an extent – it's how they kill their prey. But that venom – enough to kill a fly and inflict pain or irritation on humans – isn't really a reason to fear them, because rare is the British spider that has the ability to deliver an effective bite. Our skin is thick and rough, and their jaws just aren't up to it.*

But the tabloids don't care about that.

* The same, incidentally, goes for the house centipede (*Scutigera coleoptrata*), bless its dozens of cotton socks. It's potentially alarming in appearance, what with all the legs and everything, and surprisingly speedy if given its head, but its jaws aren't big enough to exert a grip on any part of the human anatomy.

The true narrative – British spiders are harmless to humans and in fact play an important role in healthy local ecosystems – is far less dramatic than the false one: THE POISONOUS BASTARDS ARE COMING TO GET YOU CLOSE EVERYTHING DOWN AND RUN AROUND IN A BLIND PANIC THEN KILL THEM SMASH THEM.

At a time when we need, more than ever, to be aware of the natural world and understand our role in it, it's a depressing state of affairs.

But if spiders are maligned, then spare a thought for the common wasp (*Vespula vulgaris*). Target of many a rolled-up newspaper or picnic waft, they are singularly unloved in the insect world. And all because they're attracted to your apple and blackberry tartlet.

Well, not just that. The stinging doesn't help. A wasp sting is painful, and often seems unprovoked. We don't regard flapping a pest off the aforementioned tartlet as a provocative act – we just want to protect our food. But all the wasp sees is an unidentified predator attacking it for no good reason. This is just one of many examples of how wildly human and vespine perspectives on life diverge.

It's different with bees. Bees, with their waggle dances, honey and pollinating, are regarded as benevolent. Our affection for them is even reflected in the name many people use as a blanket term: bumblebee.* That word, 'bumble', is redolent of a sort of well-meaning clumsiness, almost as if they've survived this long on the planet by mistake, wandering around the garden for 120 million years looking for their glasses without realising they're on their head.

* There are twenty-four species of bumblebee in Britain, of which eight are widespread.

Compare our contrasting reactions to bees and wasps: when we find a bee in the house we usher it out with solicitude and gentleness; find a wasp and we squash it.* And yet we're more likely to be seriously hurt by a honey-bee attack than by a wasp sting. If you're unlucky enough to be stung by the latter, it will hurt a bit, but swatting the insect away thereafter will be enough to send it packing, and its cohorts generally take a laissez-faire attitude to the suffering of one of their number. Take the same action for a honey-bee sting and it amounts to a call to arms. The bee will die, leaving its sting in your flesh and releasing a scent that the rest of the hive recognises as a signal to attack the enemy.

That's you.

And at that point things can get serious, although it's fair to say the risks are higher if you're a person who spends a lot of time with bees.

Nevertheless, the wasp persecution persists, based on a widespread belief that they contribute nothing to society. Because they don't make honey (a failing common to many other species) or pollinate (not strictly true – there are several species of pollinating wasp), we regard them as useless layabouts – especially the ones you find drunkenly crawling across the sitting-room carpet at the end of the season. So it's good to know that many species of social wasp are useful in the garden, predating smaller insects often regarded as pests. Fling that in the face of any wasp-haters you meet.

Any resentment we harbour for wasps is multiplied for hornets, simply on the basis of size. The relatively recent incursion of the darker Asian hornets – a predatory species capable of wiping out

* You don't. You're nice. I mean people in general.

a whole bee colony single-handedly – has muddied the waters, but the fact remains that the vast majority of these much maligned species present far less danger to humans than we've been brought up to believe.

If we find it difficult to dredge up sympathy for wasps and hornets, surely we can find it within ourselves to pity the poor hoverflies (*Volucella zonaria*, to name but one)? There might be sound evolutionary reasons for their resemblance to wasps* – it's a handy trick to avoid predation – but it seems a less wise move when the sight of one leads a skittish human to reach for the rolled-up magazine at the merest glimpse. Never mind that they're completely harmless – and, as outdoor insects, trying to get out of the house, not into it – squish them we must.

Similarly, the woodlouse (*Oniscus asellus*) can find itself trapped in an environment not to its liking. You're most likely to find these lovable little land-dwelling crustaceans under a log, or somewhere else that provides the moisture they need to survive. But they're not equipped to deal with extremes of either wet or dry, so heavy rainfall will see them scuttling towards the shelter of human habitation, where all too often they succumb to dehydration brought on by excessive warmth. So if you see a woodlouse indoors, the most helpful thing you can do is to usher it outdoors with a gentle but unyielding hand.

This benevolent St Francis attitude is all very well, but sometimes it makes sense to harden the heart a little. When animals get into your food supply, there's Health and Safety to consider, not to mention the small matter of survival.

* The same applies, incidentally, to a fair few species of moth, too.

While I take the presence of a caterpillar in the lettuce as a sign that my food has been grown in an environmentally sensitive way (not to mention the opportunity to enhance my diet with a bit of free protein), I would baulk at welcoming a bacon beetle (*Dermestes lardarius*) into the provisions cupboard. These creatures (and others of the larder beetle family) were quite the thing back in the day, before the advent of modern packaging, their determination to get at the food enabling them to burrow their way through wood. And you probably don't want me to tell you about the grain weevil's (*Sitophilus granarius*) habit of drilling a hole into grain seeds and laying its eggs in the hole, from where the invisible carnage can take root.

But if we're less likely, in these days of enhanced kitchen hygiene, to open a cupboard and find an infestation in our food supply, there are still plenty of tiny chompers out there prepared to make our lives a misery in their own special way.

Take, if you will, *Tineola bisselliella*. You might know it better as the clothes moth.*

And already I can hear the swearing.

Because no matter how attuned to nature we are, no matter how much we believe that all life is equal, no matter how diligently we adhere to the principles that underpin an ethical and cruelty-free lifestyle, clothes moths are bastards and there's an end to it.

You might defend *Tineola bisselliella,* as I occasionally have, with the observation that in chomping its way through your favourite jumper, this moth is merely fulfilling its role in a complex and multi-layered ecological system, and that surely we can set aside our

* It has a cousin, the case-bearing clothes moth (*Tinea pellionella*) – different name, same effect.

personal despair at the loss of a much loved article of clothing to revel in their position as a fascinating example of the biodiversity that has evolved over 3.5 billion years on this planet – but that argument is easily countered by the pithy observation that sod that, I paid eighty quid for that sweater.

Strictly speaking we shouldn't direct our wrath towards the moth itself. Adult clothes moths are merely shagging machines – their only role in their short life is to reproduce. They don't need to eat – they're still full from that sock they gorged on as a larva, and, besides, their mouths have atrophied – and once their reproductive business is done, they just sit on your wall and wait to die. The damage is done by the larvae, for whom ideal conditions include warmth, a touch of moisture, and natural fibres infused with organic fluids such as human sweat. No wonder they like your jumpers. It doesn't help our anti-clothes-moth agenda that we make it easy for them by keeping our houses at the perfect temperature for an accelerated life cycle. And with egg clusters of up to two hundred, an ability to spin mats under which they can hide to avoid detection, and an un-moth-like preference for the dark and shady, their abundance in modern centrally heated homes and resistance to human intervention is no surprise.

As with any insect designated as a pest, you can take action to get rid of it. Chemicals* are the preferred option for many people, but other, gentler methods include freezing, asphyxiation, burning, moving to Antarctica, or never wearing any clothes ever again. Sticky pheromone traps pull off a neat little quantum trick by accumulating

* None of which are in any way detrimental to human health. Nope, definitely not.

a decent collection of dead moths while not apparently having any impact on the level of infestation. Old-school remedies such as lavender bags and red-cedar balls make your drawers smell nice but that's about it, the levels of concentration required to have an effect being much higher than those admittedly attractive options are able to offer.

Taken out of the context of causing widespread, albeit low-level, human misery, *Tineola bisselliella* is rather attractive in its own way. It's a member of the subcategory 'micromoths',* so is what professional lepidopterists call 'pretty small' (about 6 mm). It catches the eye as a buff-coloured blemish on the skirting board, but if the light falls on it in a certain way it shows up an ochreous lustre that in other circumstances (in the plumage of a golden plover, say) would induce a small sigh of satisfaction. But, as already discussed, the sighting of a clothes moth isn't 'other circumstances', and even the most ardent animal-lover's tolerance will surely be stretched to the limit by their predations.

The common theme here is the dividing line between humans and nature. We build buffer zones around us, permitting access only to the chosen few. The home is a place for humans, not nature. It's our sacred space – intruders not welcome.† It would be great if we could brush away all our prejudices and misunderstandings, and see

* Don't be gulled into thinking this necessarily means they're smaller than 'normal' moths. Most are, but some aren't.

† Except when it comes to miniature lions and wolves – we'll let any number of those into our lives. We love cats and dogs to the point of species-wide self-delusion. Humans congratulate themselves on their relationship with cats and dogs, imagining fondly that they have domesticated them, when to any objective observer the truth is quite clearly the other way round.

things as if for the first time, with a sense of curiosity and wonder. What is this and why is it and what does it do?

Difficult as it can be to remember when your sweater's been chomped or your arm's been stung or there's a writhing mass of ick in the corner of the bedroom, these creatures aren't being this way because they're vindictive towards humans; they're just trying to get along in a harsh and cruel world. And that's something I think we can all identify with.

2

A GARDEN IS A LOVESOME THING

In which our author ventures outdoors, gets his hands dirty, extols the virtues of soil, and explores the wriggly predilections of a certain Mr Darwin

'No garden, however small, should contain less than two acres of rough woodland.'

It's a fine ambition, and one to which I'm sure we all aspire, but the words of Nathaniel, 1st Baron Rothschild, whose garden at Tring Park contained rather more than that, will have a hollow ring to anyone looking out of their back window over a 10-square-metre area of mossy flagstones and a brick wall.

While it's easy to mock the Noble Lord's words – and they may well be apocryphal in any case – they bring into focus a problem facing any gardener: how do you manage the resources at your disposal? Give a hundred people that 10-square-metre area, and you'll get a hundred different gardens. Some will dig up the flagstones and conjure a horticultural miracle, making full use of every square centimetre; others might fill it with pots from the local garden centre and watch forlornly as they all die for want of care and attention; some might go for the gnome option;* others still might make a desultory attempt to enliven the place and then lose the will completely, racked with indecision, condemning it to a perpetual fate as home for a corner of Busy Lizzies and a dying basil plant from Sainsbury's.

But, regardless of size, the garden – from window-ledge herb box to statuary-strewn parkland – represents an opportunity for humans to make their mark on an outside space. And as anyone who has ever had a garden will attest, it's also an opportunity not only to welcome nature in, to nurture and tend to it, but also to keep it at bay with every means at our disposal.

We love nature, but only if it signs our Terms and Conditions.

I am a lifelong non-gardener. Not only that, I am the worst kind: one married to a gardening professional. A little knowledge is a dangerous thing, but not half as dangerous as two decades being spoonfed horticultural luxury on a daily basis. I now take it for granted that at any time of year I can stroll to the bottom of the garden and be treated to a feast of sensory stimulation, whether it's the sap-burst of burgeoning spring, the bright colours and heady scents of flowering plants in summer, the melancholy decay of autumn or the subtle variety of greenery that forms the backdrop to a winter garden as

* There is a front garden not far from me that adopts this policy. When I walk past it I'm hard pushed not to gawp.

plants hunker down and take a rest before the season of growth begins again.

It's with the spirit of Lord Rothschild in mind, and with due trepidation, that I describe our garden as 'small'. Compared with the 300-odd acres Rothschild had at his disposal, this description is certainly accurate; next to the aforementioned mossy flagstones, it's Avalon. It measures, I've just worked out, 33 metres by 7. Like many city gardens it's the width of the house, and backs on to a similar garden in a similar street. Bounded by wooden-panel fencing on each side and a bramble-smothered wall at the end, it's much like many urban gardens: terrace at the top, shed at the bottom, space in between to do with as you will. Our neighbours on one side have patchy grass, a trampoline, a swing, a table, a pile of bricks and two inexplicable car tyres; on the other side, the space is notable for decking, shrubs, a large pergola and good intentions. Incursions into other people's territory are rare: the odd football flying over the fence, an overhanging tree or two,* but mostly we keep ourselves to ourselves in true London fashion.†

When we moved in, ten years ago, we could have done anything with our space. What was left by the previous tenants was nothing to write home about, but as this was now our home there would have been little point in doing so anyway. The world, or a small part of it, was our oyster. We could have made a herb garden as might

* The good-intentions neighbours have a lovely tree that overhangs our fence and acts as a home base for goldfinches and blue tits between raids on our bird feeders, so I'm definitely not complaining.

† The guest of a former neighbour once threw a lit cigarette butt over the fence. It landed at my feet as I was eating lunch. They got it back immediately, accompanied by a complementary bark of rage.

have been found in a thirteenth-century monastery; a miniaturised emulation of the parterres at Wilton House; a classic English cottage garden, flowers and shrubs and fruit and vegetables all bundled in together, overflowing with abundance. We could have constructed a model village, a crazy-golf course, or – yes, indeed – a rampant collection of plastic gnomes. We could even, as I only half jokingly suggested, have paved the whole thing with AstroTurf and installed a cricket net.

We didn't, of course. Because what we needed at the time was a lawn. A good, hard-wearing London lawn, for the playing of ball games and general small-child-based rambunctiousness.

In this we weren't unusual. Few things encapsulate the British approach to gardens like the lawn.* Historically, we've loved them, from the smooth, lush bowling greens that became fashionable in the thirteenth century to the archetypal suburban lawn of the second half of the twentieth, mown to within an inch of its life and denuded of anything extraneous,† even down to the merest hint of moss or clover, and all in an effort to show the Joneses what's bloody what. The close-cut lawn is the apogee of our obsession with everything being neat and tidy, including nature. Get your hair cut, tuck your shirt in, shine your shoes, prune the roses, mow the lawn – the vicar's coming to tea.

* The word comes from the French *laund*, meaning 'an open space among woods'.

† According to the precepts laid down in *The Lawn Expert* by D. G. Hessayon, whose series covering every conceivable aspect of domestic gardening has now sold more than 60 million copies. I bet you have a copy of at least one of them on your shelves – whether or not you ever consult it is another matter entirely.

But needs evolve, and so do gardens. Rambunctiousness turns into something slower and gentler. And so the flower beds nibbled into the lawn, plants of various sizes and colours appeared, and what was once a playground is now a haven of peace and horticultural abundance.

I had little to do with this process, merely nodding with a mixture of seigneurial approval and admiration as Tessa made yet another suggestion that would make the garden seem, Tardis-like, larger than it was – not to mention more abundant and fertile – and then proceeded to implement it.

If we don't have the required two acres of rough woodland, there are remnants of the surrounding area's past as the gardens of a local manor house, and there remain half a dozen or so well-established large trees in the immediate vicinity – oaks, a couple of planes, an arresting cedar of Lebanon two doors down – to give interest to the skyline. They also house a variety of birds, most of which, at one time or another, have made forays to the feeders on our terrace.

While these trees do their bit in lending the backdrop to our garden a bit of variety in a suburban landscape, it's difficult to ignore that at the heart of the whole set-up – and the ugliest element of it – are the fences marking the boundaries to our property. They're a constant reminder that the root of the word 'garden' is the early Indo-European word *ghordos*, meaning 'enclosure'.* Clear straight lines, delineating what's what, where's where, and whose is whose. 'This is mine, that's yours' – but also: 'Out there is wilderness,

* Pleasingly, the word 'paradise' also derives from a word (*pairidaeza* in Persian) meaning 'enclosure'. So a paradise garden is an 'enclosure enclosure'.

nature in the raw; in here – this is for humans.'[*] But nature abhors a straight line, and everywhere there's evidence of its unerring instinct for encroachment – tendrils of bindweed creeping under the fence, wisteria[†] snaking over it, ivy engulfing everything.

If wisteria is welcomed for its fetching annual display – light lilac flowers hanging in drooping bunches from gnarly branches against a whitewashed wall being almost obligatory for a certain kind of suburban house – the other two cause more consternation. Ivy's dark and leafy attractiveness is offset by its tendency for rampant growth, giving it a reputation for attaching itself to buildings and sucking the life out of them.[‡] And while my innocent reaction to the elegant white trumpet-shaped flower of *Convolvulus arvensis* (bindweed, to give its common name[§]) has always been 'Oh look how pretty', the gardener's is 'Fetch me the flame-thrower, Beryl.'

* Once upon a time, back in the day, some gardens would have been sacred groves, devoted to the gods, and therefore not a place for humans. They're not that fashionable in this day and age.

† It's named after Dr Caspar Wistar, an American physician, yet the official spelling remains 'wisteria'. Other plants to suffer evolution of either spelling or pronunciation include fuchsia (named after sixteenth-century German botanist Leonhart Fuchs but routinely pronounced 'fyoosher') and dahlia (named after eighteenth-century Swedish botanist Anders Dahl but pronounced – in Britain, at least – 'day-lee-a'). Yes, you're right – I should let it go.

‡ It doesn't do this. It's also unfortunate that, just as ivy's getting going with the flowers in the autumn, we have a habit of cutting it back. This is bad news for the ivy bee, a relatively recent arrival to this country.

§ Its roots can go to a depth of twenty feet, which does I suppose explain in part the gardener's aversion to it. That and its habit of twining itself round other things and gradually smothering them.

This separation between plant and weed, acceptable and unacceptable, is bewildering to the uninitiated, but second nature to gardeners. It's also embedded in the history of gardening. When the Romans* found their way to Britain a couple of millennia ago, they changed the face of gardening along with everything else. Their introduction of an abundance of plants – among them quince, mulberry, leeks, turnips, kale, asparagus, parsley, dill, marjoram and much much more – was accompanied by pesticidal products and techniques: tar, bitumen, fumigation. And the struggle has continued ever since. This taming of nature has varied in degree down the ages, but at the heart of it is the imposition of human influence on the wild – the illusion of balance achieved only by constant efforts to keep nature in check. From the grandest excesses of the formal gardens of stately homes – with rampant statuary, fountains, topiary and grottoes – to the smallest urban 'pocket handkerchief', the garden has always been an extension of our indoor sacred space: access denied to the interloper, be it mildew or mare's tail, squirrel or slug.

The impossible balance – being simultaneously both with and against nature – has always been at the heart of gardening. But there are other underlying threads. The garden is a form of self-expression – this is who I am, what I like, how I see the world – but it's also a reflection of the prevailing concerns and character of the society of the day. We need only to look to the twentieth century to see the rhythms of hardship, recovery and comfort find expression in gardening fashions. Both world wars saw a natural intensification of subsistence gardening at all levels of society – George V grew potatoes in the flower beds at Buckingham Palace in the First World

* Of course it was the Romans. Who else was it going to be?

War, and the slogan 'Dig for Victory!'* led to massive use of both private and public land for the growth of vegetables in the Second. And when the fighting was, at least temporarily, done with, people looked to their outside spaces for solace and pleasure. In both the 1920s and the 1950s, home ownership blossomed, and with this growth came a renewal of interest in the garden as a place of pleasure, pride and – once the lawn was mown, slug pellets put down, fruit trees sprayed, weeds eradicated, and a million other tiny niggles identified and dealt with – relaxation.

The niggles are, naturally, part of the gardener's stock-in-trade. Their work is never done. 'Plan your garden as if you will live forever', as the old saying has it – but just be aware that most of forever will be spent rooting out dandelions, staking the apple tree, and wondering whether the *Lythrum salicaria*† wouldn't do better if it was moved over there where the ground is boggier, ooh and how about putting the *Lamprocapnos spectabilis*‡ over there to give it a bit of shade in the spring, what do you think? Even maintaining the status quo requires constant attention. To leave a garden untended, even for a short time, is to understand what nature can do by itself, and if you're going to meet it halfway you need to be prepared for it to overrun you the moment your back is turned.

But it's that quest for perfection, even while knowing such perfection is unattainable, that keeps the gardener going. Every time you put something in the ground you don't know if it'll come up

* Taken from an *Evening Standard* leader, it quickly replaced the original, relatively prosaic government slogan, 'Grow More Food'.

† Purple loosestrife, so I'm told.

‡ Bleeding heart, apparently. You don't think I'm the one who knows this stuff, do you?

or run rampant or be chomped by slugs or be engulfed by a larger, more prolific plant behind it. So you read and you learn and you get plants and some will thrive and others won't, and then you ask yourself why and get more plants and put them in different places and gradually you make sense of it all without ever quite understanding exactly why certain things work and others don't.

Gardening, like so many things, can easily be seen as a metaphor for life.

If my experience of gardens is that of a gardenee rather than a gardener, it's not that I can't see the attraction.

There's the satisfaction gained from nurturing a plant: the ritual of sowing a seed in a pot, tamping down the compost as if putting a newborn to bed; the thrill of seeing that first green speck nudge aside the soil, all promise and future; the enduring glow brought on by its slow and steady growth, and then the giddy elation of seeing it flower or fruit, or just the quiet triumph engendered by the simple fact of its continued survival. All of these feelings are valid, even if such success was merely a happy accident of positioning, timing and the confluence of horticultural circumstances over which you have no real control. It's your plant; you gave it life, so you get to reap the reward.

There's also the deeper, less immediately tangible satisfaction of contact with larger, invisible rhythms. A garden marks the seasons, reminds us of the cycles of the year, the slow changes that cannot be rushed. The innate appreciation of these rhythms might be why I have never met an impatient gardener. They persevere, taking setbacks with a rueful laugh, and then they wonder how they can do it better next time.

And while they lavish love and energy on the garden, it gives it right back, whether in the form of flowers or food or the dappled shade cast on a swing chair by a Japanese maple. It's a two-way relationship: we tend the garden, it tends us, from the wispy fragrance of jasmine flowers to the clods of earth shaken off a recently dug potato. And none of it – from *Alchemilla mollis* to *Zinnia angustifolia* – is anything without one vital ingredient: soil.*

Give a gardener a brisk cold day, a pair of wellingtons, some twine, a sharp pair of secateurs, the heft of a really good hoe, a seed catalogue, a dibber, a trug full of raspberries, a capacious water butt,† bees thronging on lavender, a compost heap, a shed to potter in, a good solid wheelbarrow, a bag of mulch or the fecund ripeness of well-rotted manure, and they'll be happy enough.

But if you want to see a gardener glow, just say the word 'soil'. Something happens to them – something fundamental. An infusion of warmth, a deep feeding of the soul.

If they're the talkative type, you won't be able to stop them. There'll be talk of loam and nutrients, of friability and pH, of waterlogging and compaction, and of chalk and peat and sand and silt and clay.‡ There will be words like 'humus' and 'black gold'. There might be a fifteen-minute discourse on the feel of soil in the hand,

* Yes, I know – hydroponics. Maybe allow me a sliver of rhetoric, however vague and inaccurate, to round the section off?

† Or possibly, if my experience is anything to go by, two.

‡ This last element is especially true of London gardeners, who can talk of little else.

the dark crumble of it as it filters through your fingers, the particular soily smell that appeals to human senses in subtle, primeval ways.

If they know you well enough, they might even try you with the word 'tilth'.*

This veneration of the soil has a spiritual element. It's not just about growing pretty flowers, although that might be incentive enough for many. You won't find a gardener who isn't at their happiest with at least a bit of dirt under their fingernails, and this connection with the earth is deep and fundamental, almost as if they yearn to sink a tap root into it, to draw life from its nutrients and become one with it. And that contact with the soil – even the simplest act of hollowing out a hole in a flower bed, inserting a plant and packing the earth around its roots – engenders feelings of well-being, often visibly and immediately.

But as any good gardener knows, dirt isn't just dirt. There's a world of life underfoot: vital, invisible, and extremely hungry.

First, take your underpants. Any brand will do, so long as they're made of cotton. Now bury them somewhere in the garden. It's wise to leave a corner sticking up out of the soil. That way, when you come back to them in a few weeks to see how they're doing, you won't spend half an hour digging up the petunias in a vain search for a pair of rotted undies.

* Which derives, I discover, from the verb 'till' – one of those 'obvious when you know it' connections. Disappointingly, the same connection can't be made between 'fill' and 'filth'.

In those few weeks a miracle will have taken place. The scale of the miracle will depend on the health of your soil. The larger its microbial diversity, the less there will be of your underpants when you retrieve them.

This is a good thing.

In every little patch of earth live a million strange and tiny beings. From the tidgiest bacteria and actinomycetes, up through the rotifers and nematodes and minuscule flatworms, then into the realm of the barely visible – springtails, potworms, fly larvae, beetle larvae, centipedes, millipedes, those little snails you occasionally come across that look like a tiny pebble or a piece of grit but it's only when you balance them on your fingertip that you see the fine coil of the pattern on their shell and you start thinking about the variety of life on earth and evolution and fractals and the mathematics of nature and that article you read about how the Fibonacci sequence is all around us* and before long you're lost in a fog of wonder about, well, all of everything.

Oh, and slugs, too. Yuk.

Or maybe not yuk, if you're of a certain inclination. It's true that from the human perspective slugs don't seem to have much going for them. They eat our crops and leave slimy trails all over the place – said slime being notable for its indelibility. But this is to dismiss the benefits they bring as both consumers of decomposing vegetation and valuable food sources for thrushes and beetles and ants and hedgehogs and any number of the aforementioned miniature beasties.

* If you don't believe me, look at a pineapple.

The same, by the way, goes for snails. We're much fonder of snails, partly because of the fractal spiral thing, but also because a snail carries its house on its back, a habit we equate, for some reason, with a sort of plucky self-sufficiency and resourcefulness. As a measure of how we view the two genera, look at how we anthropomorphise them. Brian, the endearingly dim-witted character in *The Magic Roundabout,* was a snail; Jabba the Hut was, if not an actual slug, the embodiment of everything we hate about them.

Gardeners, of course, loathe both slugs and snails* with the heat of a thousand fiery suns. If the word 'soil' reveals a gardener's deeply compassionate and human side, just try them with the word 'slug' and watch as the demons take possession, transforming them into sadistic maniacs in the grip of an unquenchable bloodlust.

But I have strayed from the main point – namely, those microscopic beasts that have chomped your underpants. Most of the grunt work in that department is done by the smaller creatures, the bacteria and nematodes and suchlike. Your underpants contain cellulose, which is catnip to the little wrigglers. And as they work their way through the material, their excretions serve only to enrich the soil. Even better, there's increasing evidence that the benefits of getting your hands dirty aren't just woolly 'Oh isn't it lovely to get some fresh air and muck around in dirt' sentiment, but actually have a scientific basis. Contact with the common bacteria *Mycobacterium vaccae,*† to pick just one example, triggers the release of serotonin, which, as I'm sure we all know by now, is a Good Thing.

* But not, interestingly, puppy-dogs' tails.

† It's related to tuberculosis, but don't let that alarm you – it's one of the good guys.

So it's clear that going outside and rootling around in the dirt is a beneficial and healthy thing to do. Throw in the stimulus of colour and scent and the general sense of well-being engendered by spending time outdoors, and it's almost enough to make me take up gardening.

But in the list of soil-dwelling organisms above, I made one deliberate omission. Leaving the best till last, you could say. Because if all those organisms are doing excellent things to the soil, they pale into insignificance next to the mighty earthworm.

And mention of earthworms leads, as night follows day, to an undeniably Great Man.

By all means, *On the Origin of Species by Means of Natural Selection, or the Preservation of Favoured Races in the Struggle for Life*;* certainly, *The Descent of Man, and Selection in Relation to Sex*; naturally, the finches and the climbing plants and the barnacles and the expression of emotion on the human face and the coral reefs and the orchids and the insectivorous plants and the fundamental recalibration of everything humans had ever thought about themselves in relation to all other life and the history of the planet.

All of those things, obviously.

But how intriguing and charming that the last work of this extraordinary life – a project that occupied Charles Darwin, on and off, for forty years – should be about earthworms. His own description of *The Formation of Vegetable Mould through the Action of Worms, with*

* We must give it its proper title, I feel – he took the trouble to write it out in full, so it's only polite to reciprocate.

Observations on their Habits[*] is rather touching: 'This is a subject of but small importance; and I know not whether it will interest any readers, but it has interested me.'

That mild-mannered understatement, 'it has interested me', conceals much. Because when a subject interested Charles Darwin, no aspect of it went unexamined. And it's somehow fitting that, having produced the most influential book of his (or any other) era, he chose to see out his days not resting on his laurels or exploiting his fame for further gain, but in pursuing his lifelong obsession with understanding how it all works.

And by 'it all' I naturally mean 'life on earth'.

Earthworms aren't, on the face of it, a promising subject, and would have seemed even less so to the Victorian reader. Despite the success and acceptance of *Origin*,[†] people still had[‡] entrenched ideas about the superiority of mankind over the rest of the animal kingdom. And worms, creatures of the dirt, would have been regarded as the lowest of the low. They still are in some quarters, I suspect, despite all we know about their importance.

Darwin's dogged study of earthworms,[§] and the conclusions he reached about their contribution to the workings of the world, encompassed every aspect of their lives. As part of his relentless quest to uncover their mysteries, the earthworms of Down House were subjected to every test he could think of. He measured their

[*] Snappy subject, snappy title.

[†] Having done my bit for the full title, I'm now allowing myself extreme abbreviation.

[‡] And, sad to say, often still have.

[§] '. . . unsung creature which, in its untold millions, transformed the land as the coral polyps did the tropical sea'.

sensitivity to heat and cold, light and dark; their powers of suction and traction; what they did to the soil and how; how much earth they moved, where they put it, and how long it took;* their potential for intelligence; their role in levelling ground and burying ancient buildings; their senses of smell, taste and hearing – this latter manifesting itself in their reactions to having the bassoon played at them, a scenario that conjures, more than any of the other studies, irresistible and inadvertently comic images.

Crucially, he went into these studies without any preconceptions. He merely did what he had always done: assembled the data, analysed it, came to conclusions.

It's that last bit that's deceptively simple. Anyone, given patience and time, can collect and analyse data – it was in the joining of the dots that his genius lay, as well as his underrated ability to express these findings in language easily understandable to the layperson. At the heart of it all was a simple truth: he loved looking at things. Not just looking for looking's sake, but looking at, through and beyond things to find the patterns and processes that drove and linked them. It was slow, relentless work, with no guarantee of results.

And here's the thing. He did it over not just years but decades. He studied and studied, observing in a way that few before or since have achieved. Thousands and thousands of tiny dots on an infinitely large piece of paper, joined by the power of perceptive observation and a rare capacity for deep and logical thought.

His doggedly unromantic approach – painstaking, going over the figurative and literal ground inch by inch – was, and is, at odds

* A modern estimate has it that earthworms turn over the equivalent of all the soil on earth to a depth of one inch every ten years.

with the way many people see and describe nature. Which isn't to say he didn't approach it with a due sense of awe and wonder. On the contrary. But by revealing the true nature of things, he exposed their magic in a different way.

It's easy to forget, too, that when he writes about nature, he includes us. We're not at the top, as had previously been assumed, the pinnacle of God's creation. We're no different from the lowly worm or any of the others: naturally predisposed to survival, slave to evolution, the fundamental objective of our existence simply and solely to make it through to the next generation.

Yes, we have enlarged brains and opposable thumbs and any number of things that have enabled us, for better or for worse, to dominate the planet in the extraordinarily short time we've been on it. But that doesn't make us better than a nematode or a horsefly or a giraffe or *Tineola bisselliella*. Just different. That's the whole point, but one that still, sadly, seems to be lost on a large portion of humanity.

As with *On the Origin of Species by Means of Natural Selection, or the Preservation of Favoured Races in the Struggle for Life,** Darwin's work on earthworms wasn't all new – although he paved the way for a new approach to the subject, he was building on the achievements of others in the field. And there was no controversy attached to it, no overturning of entrenched and fundamental ideas. But in its own way, because it focused on an aspect of the natural world that most people ignored, even in that time of discovery, the book was just as radical.

Its argument is simple: worms are amazing and important, and these are the reasons why. And as he leads you through those

* Back to the full title, just this once.

reasons, with all the evidence painstakingly laid out in prose and table form,* he leaves you with little choice but to agree, and not just to nod along, but to share his enthusiasm.

At the conclusion of the book, Darwin writes this: 'When we behold a wide, turf-covered expanse, we should remember that its smoothness, on which so much of its beauty depends, is mainly due to all the inequalities having been slowly levelled by worms … It may be doubted whether there are many other animals which have played so important a part in the history of the world, as have these lowly organised creatures.'

How elegant, how appealing to the Victorian sensibility, relating a potentially obscure subject to something commonplace like a lawn, something the reader would hold dear. He's saying, 'This sciencey thing is a homely thing too, something we can all understand.' Few have looked closer or better than Darwin; with those words he encourages us to emulate him, even if in a small way, and to remember the importance of things we all too easily dismiss. It almost makes you want to stand up and cheer.

And so, in honour of both the great man and the mighty worm, I go to Down House, just outside the small village of Downe† in Kent, to look at his lawn.

I could, and probably should, walk there. It's eleven miles, door to door. Nothing I can't manage, but time-consuming. So I compromise: train to Beckenham, then walk, a schedule that should deliver me to Darwin's front door just on opening time. It's a lovely day for

* The tables, I have to admit, do sometimes come as a bit of light relief.

† The spelling of the village's name was precipitated by ongoing confusion with County Down in Northern Ireland. The Darwins eschewed the extra 'e'.

it, and what better way to spend it than to stroll along country lanes enjoying the delights of that nightmare of the spoonerism-prone newsreader, the Kent countryside?

Several better ways, as it turns out. There's no doubt that the scenery is attractive, but the lanes around Downe have not evolved to favour the pedestrian. Cars and vans pass at speed, showing as much consideration for the benighted walker as a song thrush does for a snail. After a while it occurs to me that it might just be easier if I complete the walk sideways, pressed against the brambles on the steep bank lining the road, better to accommodate the vehicles that are in such a hurry to get to wherever it is they're going.

It's easy to think of all the things that have changed since Darwin's time, but on the approach to the house, and on arrival there, I'm more struck by all the things that are the same. Take away the Ocado vans and Ford Mondeos and the light aircraft occasionally taking off from nearby Biggin Hill airfield, and there's a timeless quality about this part of the world. It's even possible, albeit with a bit of squinting, to picture the surrounding fields as they would have been when the Darwins moved here in 1842.

This timelessness is only enhanced on entering the house, which has been lovingly restored by English Heritage with many of Darwin's own furnishings and artefacts, as well as faithful period decoration. Exercise a bit of imagination and you can plonk yourself in the middle of family meals in the dining room, watch Charles and Emma conducting their daily backgammon duel in the sitting room, or imagine son Francis playing the bassoon to a bell jar of earthworms while father Charles takes diligent notes on their reaction.

I'm taken by all the details of the house, and the richness of the exhibition upstairs, which brings vividly to life Darwin's many achievements, from the voyage on *The Beagle* to the groundbreaking work on barnacles and beyond. But it's only when I enter the study that it hits me.

This is where it happened. Here. In this room. All that thought and work and insight, right here where I'm standing. That rectangular Pembroke table, laden with scientific instruments, notebooks and papers; that horsehair armchair, raised on castors for ease of movement about the room; those glass-stoppered bottles and pill-boxes with powders and tinctures and specimens; the microscope, the brush, the rock samples, the bones – all his, all in front of me, a monument to quiet industry.

It's easy to fetishise these things – 'I say, look at this, a piece of dirt once trodden in by the great man himself!' – but there's something about this room that facilitates the connection between the visitor and the genius. Something about the darkness of it, sunlight trickling through the wooden shutters for only a couple of hours each day. Something about the way it's been preserved yet somehow kept alive. Something, dare I say it, metaphysical.

I spend half an hour in there, allowing this atmosphere to settle on me, in no hurry to move on. There are no interruptions, no puncturing of the bubble. I have, for just a moment, the fanciful thought that I've stumbled through some sort of portal, and that any moment now he will pad back into the room, returning from some household errand to the serious matter of work. It's an impression only strengthened by the sound of a man's voice murmuring outside in the corridor. And then he comes in, the first person I've seen for half an hour, and it's one of the group that was in the entrance hall when I arrived, a modern person with a modern phone and a modern leather bag, barging into my Victorian time travel and talking over his shoulder to someone else out in the corridor.

'Oh, we've done this room already. Let's go to the cafe.'

He turns and leaves. I return reluctantly to the real world, and head out to the garden.

Outside, the spell descends again.

I'm on the lawn, looking back across it at the house. Just a few yards away, across a gravel path, are the greenhouses, where Darwin conducted experiments on orchids and climbing plants and insectivorous plants and honestly it's exhausting just thinking about the scope of his lifetime's work.

The greenhouses are still equipped with the heating pipes Darwin installed to keep the tropical-plant specimens he was studying at a suitable temperature. They're tempting, but I leave them for the moment. I'm here to look at a circular stone slab embedded in the grass. Its purpose would have been obscure to many, but to the Darwins it was an important long-term monitoring device. They called it the 'worm stone', and along with a dinky piece of copper-and-wood apparatus now on display in a glass case in the house, it was used to measure the amount of soil displaced by earthworms.*

The location of this experiment is a little corner underneath a Spanish chestnut, some distance from the house. On a day like today, warm in the early autumn sun, a lawn like this is a positive invitation to lay a tartan rug on the ground, tuck into genteel cucumber sandwiches and Victoria sponge cake, and then doze quietly as

* Approximately 2 mm a year, in case you're wondering.

the children charge around playing some invented game, the rules known only to the participants but no doubt involving a lot of running and shouting.

But I have one more part of the Darwin estate to visit, so, resisting the temptation to dig down into the lawn with my bare hands in the hope of extracting an earthworm, I walk down the gravel path, through the abundant and extensive kitchen garden, turn left through the gate in the wall, and into the woods. I'm following a circular route Darwin took every afternoon, allowing the researches conducted in the morning to percolate through his brain and join the mulch that fuels all sorts of creativity and advancement.

It's easy to think that genius comes in a flash – *Eureka!* – but the truth is far more mundane. The moment is typically preceded by immense quantities of thought, hard work and confusion. And often the flash of inspiration strikes when the mind is allowed to roam unfettered, rather than having to focus on the work at hand. It's no coincidence that Archimedes had his *Eureka* moment in the bath – baths are breeding grounds for inspiration. And so are woodland walks.

This is a good one, as they go. Peaceful, solitary. Probably about two acres, give or take, of rough woodland,* rented by Darwin a few years after he moved to Down House, and relatively unchanged since, with the most obvious exception of the odd laminated activity card hanging from a piece of string to encourage the young to explore.

The relative hubbub of the house, with its tour group and cafe, feels a mile away, and I take it slowly, consciously tamping down my natural inclination to stride out at a pulse-raising pace. This walk was

* Lord Rothschild would approve.

known to the family as the sandwalk, because of the red sand that held together the gravel on the path. He would do five laps, flicking aside one of five flints by an ash tree at the entrance point for each completed lap.

There are no flints that I can find, but I manage to keep track of my three laps in my head, and as I walk I do what you do on a walk in the woods. A pleasant saunter, drinking in the tranquillity and the light and the smell of the soil, pondering things in general and Darwin in particular. It's tempting to think that his aura somehow rubs off, that simply by visiting this place we are somehow improved. But if I recognise how simplistic this is, I do at the same time feel a tiny shift in me, an unexpressed resolution to slow down, to look better at things, not to take them for granted.

As I finish the third lap, and as if inspired by my surroundings, a thought – almost a stroke of genius – flashes into my head.

Tea. Cake. Home.

Even if we can't all achieve greatness, it's important to notice and cherish the small moments of inspiration.

There is, inevitably, a shop. I buy a jar of Down House honey – made, no doubt, by the descendants of Darwin's bees – and two books: *On the Origin of Species by Means of Natural Selection, or the Preservation of Favoured Races in the Struggle for Life* and a slim volume of his autobiographical writings.

At this, the volunteer at the till gives a small nod of approval.

'Page 73. Look at page 73.'

He's friendly but firm. It would be rude not to. Quickly enough, as I read, it becomes clear to me the words he's thinking of: '. . . I can remember the very spot in the road, whilst in my carriage, when to my joy the solution occurred to me.'

I close the book and pay for it.

'I would love to know where that spot is,' he says. 'Just to be able to stand there and drink it in.'

I wonder if I passed the spot on my walk to the house that morning. It's entirely possible. Perhaps it was even where I was nearly killed by an Audi with a grudge.

'Me too,' I say. 'Me too.'

I pick up the books and the honey, and take my leave.

3

YOUR PATCH OR MINE?

In which our author makes the acquaintance of a fine tree, thrills to the return of 'his' swifts, meets dogs of varying degrees of excellence, and visits the patch of Gilbert White, the Patchmaster Extraordinary

It's a perfectly normal tree in a perfectly normal park in a perfectly normal part of town. An oak, neither massive in the way of the grand old oaks of yore nor stunted by competition from bolder, more dominant trees. It does fine, part of a haphazard line adding interest to the broad descending sweep of hillside. Another oak a couple of metres away overshadows it slightly, and a smaller lime, a little further away, breaks up the continuity. Beyond it, south London, the *ne plus ultra* of urban sprawl, the twin towers of IKEA poking up rudely from the hazy carpet of suburbia.

Norwood Grove – once the garden of the house at the top of the hill, now a public park. It's the nearest worthwhile expanse of green to our house and a frequent haunt when I feel the need for the restorative power of nature.

Pre-industry, it was part of the Great North Wood, the sprawling landscape of woodland that stretched across several miles of high ground all the way from Deptford to Selhurst. But suburbia nibbled away at it, as it tends to. It might have devoured it altogether had it not been for the power of community. A committee of local residents set up a fighting fund after the First World War to stave off developers. As a result, Norwood Grove's thirty-two rolling acres of meadow and wood were preserved in perpetuity for the public's use.

Attached to Norwood Grove is Streatham Common, with its fragment of ancient woodland, semi-formal gardens and long, grassy open slope leading down to the A23. Together they form what for want of a better word I call my 'patch'.

Talk to any birder, even a quarter-serious one like me, and it won't be long before the word 'patch' is mentioned.

'What's your patch like?'

'Had a garden warbler on my patch the other day.'

'Nothing about on my patch at this time of year.'

Your patch is your territory, where you go to watch birds on a regular basis. Mine is, by necessity, urban. It is in fact a patchwork of patches, each local green space attracting my attention in different ways, none quite getting the scrutiny it deserves: the local cemetery, full of nooks and crannies and overgrown bits, home to trusting urban foxes and winter flocks of thrushes; Tooting Common, just too far away for the casual stroll but offering more variety and interest thanks to its pond and more extensive woods; Dulwich Park, Brockwell Park, Crystal Palace Park, each with its own specific flavour and charms, each visited less often than I'd like.*

Some people's patches are glamorous – areas of great natural beauty, homes to a wealth of breeding birds in a variety of habitats, yielding counts of fifty or sixty species per visit. I'm happy in the knowledge that I'm likely to encounter no more than twenty-five, if I'm lucky, and that the most exotic visitor to the area was a hoopoe in 1886.

It's mine, and I love it.

I have my routine. To go against it feels somehow wrong. Up the hill, right at the lights, past the Lidl-that-will-never-be-built,† left and along the broad curving road until the gate.

There are distractions en route, nature finding a way even in the least hospitable places. A dunnock perched on a lamp post, raising its voice to be heard through the passing traffic; dandelions poking through between the gaps; street trees dotting the pavement, fighting the good fight all by themselves; lichens colonising the top of the old postbox on the corner – each worthy of at least a moment of my attention. And as I look at each of them, they help mute the noise of the man-made drear: concrete, tarmac, metal.

By the gate, there's a welcome party – feral pigeons, pecking busily among the debris of a fox-ravaged bin bag on the pavement. I look closely, and notice a stock dove in with them, its black-button eye setting it apart from its uncouth cousins. They're heckled by an aggrieved parakeet from the branches of a plane tree above.

* The astute reader will now be able to work out roughly where I live. Drop in for a cup of tea the next time you're in the area.

† Reader, it was built.

I open the gate and go in.

Clockwise, always clockwise. A row of hornbeams on one side, copper beech on the other, and behind them a meticulously tended bowling green. These greens, like cricket pitches and tennis courts, are largely sterile for wildlife,* but they serve a generally benign recreational human purpose, and the rigorous geometry of the mowing pattern is aesthetically pleasing in its own way.

I say 'a row of hornbeams' casually, the lobbing in of a simple fact designed to give the impression that I know what I'm on about,† but it took me ten minutes of concerted consultation with a Tree ID app on my phone to sort out exactly what it was. Unlike birds, trees are happy just to stand there while you examine them for identifying features. But despite this I struggle with them, easily confused by variations in leaf shape and bark texture. And when the app asks me if the bark is green or brown or grey, I silently bemoan the lack of a 'sludgy-browny-greeny-grey' option.

As I work my way round the grove, I note with pleasure the presence of what one birding friend calls 'the usual rubbish'. I know what he means – many birders show visible signs of excitement only at the prospect of a rarity – but beg to differ. A song thrush wrestling with a worm? Sign me up. Two wood pigeons having a bit of a fracas in the cypress twenty yards away? Bring it on. A grey squirrel bounding along a branch and springing with grace and elan across to the next tree? Yes, please, even though they're ubiquitous and widely reviled.

* Although, to judge by the ubiquity of pied wagtails at cricket grounds countrywide, they must have something going for them in the invertebrate department.

† I do know that they thrive in the clay soil that is the bane of the London gardener's life.

If the squirrel's springing with g and e across to the next t, that means it won't be terrorising the blue tits in our garden.

For all that city living has its merits, it would be unbearable without these little moments. They nourish me, and I welcome them. I welcome the hornbeams, the argumentative wood pigeons, the song thrush with its worm. And yes, I even welcome the parakeets – their excitable squawkings are simply their way of being. They are the stereotypical noisy Italian family just moved in next door in the formulaic 1980s rom-com of life.

All these and more are available to me as I skirt the grove, ducking deftly under the low-hanging branch of a wonky tree* that might otherwise deliver a numbing blow to the unwary. Down the slope to the yew tree where the goldcrest sometimes hangs out, check the starlings darting to and fro over the road, back up the gentle hill to the little copse the parakeets noisily call home, and then the gentle trudge across the grass.

It's easy to take for granted the simple benefit to be gained from grass. It's green. Greener on the other side, of course, but green nonetheless. And as numerous scientific studies have shown, green is good. The health benefits of time spent outdoors are well documented, and the Japanese habit of *shinrin-yoku,* or forest-bathing, threatens to topple *hygge* and *lagom* from their exalted positions in the competitive world of 'relaxing things people in other countries take for granted that the British suddenly get excited about'.

Grass, in particular, has its attractions beyond the merely visual. There's the smell of it when freshly mown, the feel of it underfoot on a warm spring day, the sound of it swishing in a light breeze.

* As yet unidentified – watch this space.

Whenever I see a sign saying 'PLEASE DO NOT WALK ON THE GRASS', I'm tempted to erect a twin sign next to it, bearing the question 'THEN WHAT THE BLOODY HELL IS IT FOR, YOU DRIVELLING IMBECILES?' To deprive humans of the simple pleasure of walking barefoot on grass, or lying on it and idly toying with a dandelion, or sitting cross-legged on it with a friend and talking about nothing in particular while eating ham and bread and maybe a good cheese and probably a really juicy pear now you mention it – well, it seems like an act of unthinking cruelty.

But if I'm aware, in a general way, of the health benefits and calming effect of greenery, my engagement with it is generally non-specific. Greenery is good; look at the greenery; how lovely.

Time to look a bit closer. Time to say hello to the Perfectly Normal Tree.*

I choose the Perfectly Normal Tree over the others because of its main distinguishing feature: on its bark, taking care to fill in the letters over the rough, fissured surface, someone has painted the words 'HUG ME'.

I've walked past this tree many times, often glancing towards it with an unspoken appreciation for its message of amity and peace. But, I realise, I've never looked at it. Not properly. As one whose way into nature was through birds – with their flapping and screeching and twittering and pecking and generally being eminently watchable – the idea of appreciating something so fundamentally static requires an adjustment, a change of pace. It's easy enough to thrill

* Readers of Douglas Adams will know that I have appropriated and adapted his 'Perfectly Normal Beast' for my nefarious purposes. I apologise to his memory.

to the speed of a peregrine stooping* on its prey at 200 mph, or the glorious spectacle of half a million starlings murmurating before settling down to roost. But trees just stand there, swaying occasionally.

Except, of course, that they don't.

Because the Perfectly Normal Tree is a phenomenon.

It's an oak, which as we all know represents Englishness in a way that nobody can really explain. The tree of navies past, the tree for everyone from monarch to commoner, the most English of trees. It's our national emblem, so it must be quintessentially English. But it's also significant in Bulgaria, Croatia, France, Germany and Latvia, among many others, so perhaps we should just get over ourselves.

This particular specimen won't win any prizes. Neither brazenly magnificent nor pathetically scrawny, it's an average tree. The presence of the competing lime a few yards away means it hasn't quite been able to follow its instinctive path, which is for the branches to grow out sideways – so it doesn't have that traditional rounded shape familiar from illustrations and photographs. It does have a small but discernible list, and the branches on the non-listing side are slightly scruffier than their counterparts in a way common enough in trees that have endured some hardship or are fighting for resources with competitors in an area that is in some way denuded or not in a prime location. If it weren't for the message of peace daubed on its gnarly bark, I wouldn't give it a second glance. But if I've learned anything in my few years of nature observations it's not to overlook the apparently mundane. So I hang around the tree and get to know it a bit.

* Not to be confused with a swoop, a stoop is a controlled dive performed by birds of prey in pursuit of their quarry.

Despite the invitation, I don't hug it. I'm keen on it, for sure, happy to love and appreciate it, but such a public display of affection seems, at that moment at least, a bit much. I do, however, feel it, laying my hand on the bark, probing the deep fissures in its rough, hard surface, then cast my eyes upwards towards the canopy, the branches intertwining as they reach for the sky. I contemplate its intricate geometry – as individual to each tree, no doubt, as fingerprints to humans – allowing myself to get lost in its patterns until everything is tree, branches and leaves. I'm not looking for or at anything in particular – just drinking it in, exploring the delicate lattice pattern the tree's offshoots form against a backdrop of light blue sky.

It's a slow game, at odds with the pace of modern life, inducing a palpable deceleration, a banishing of urgency. And as the tree becomes, however briefly, my world, something floats into my head – the memory of a mini-anecdote about Anton Bruckner, the nineteenth-century Austrian composer whose long and spiritual symphonies either send listeners into paroxysms of quasi-religious fervour or bore them to tears. As an old man he liked to go for long walks in the woods and count the leaves on the trees, a task as apparently fruitless and time-wasting as counting the notes in his symphonies. This story was told to me as evidence of his increasing madness. But now, standing under this oak and losing myself in its bosky embrace, it seems to me an entirely logical and sane thing to do. The people who don't do this kind of thing, who resist these moments of connection and, however unconsciously, associate with nature only in the loosest possible way, if at all – they're the mad ones.

The only drawback is that after ten minutes I have a stiff neck.

If I'm fascinated by the externals of a tree, my reaction to the workings of its innards is more akin to an explosion of awe. How many times have I walked past a copse or through a wood, or even looked out of the window into the garden without giving a second's thought to the constant, frenetic activity taking place underneath the trees' calm exterior?

We see a tree's bark, its twisted branches, fluttering leaves, flappy catkins, the satisfying slotting of an acorn in its cup. Sometimes we see the top of their roots surging up through the soil, and occasionally, when circumstances conspire to leave them exposed, we're granted a privileged glimpse of the knotted, subterranean complexity that reaches deep into the earth – further than we imagine. But for every bit of a tree we do see, there's an invisible counterpart, doing miraculous things to keep it surviving in a tough world of depredations and hostility.

While the spread of an oak's crown is a heart-warming and delightful thing, it's mirrored – doubly so, in fact – by the roots, whose underground spread far outstrips the crown's skyward reach.

And of course it's not just about the roots. The complexity and extent of a tree's relationship with its surroundings boggles the mind. Where the roots go, there too goes an astounding network of mycorrhizal* filaments, transferring water and nutrients from plant to plant, protecting them from disease by filtering out the bad stuff, fending off bacteria, and generally being a boon to the forest community. Not that the fungi don't get their reward from the bargain, gleaning whatever nourishment they need from the trees as compensation for their contribution to the well-being of their environment. And

* A word that, no matter how I write it, always seems to be spelled wrong.

then, when the time is right, they fruit, and we, on a damp, blowy autumn walk, might see them and say, 'Oh look, what a gorgeous mushroom', paying scant attention to the miraculous network of activity beneath our feet.

Speaking of miracles, think of the way a tree distributes water from roots to leaves, in defiance of gravity. We take it for granted, but the process remains more or less a mystery – whether it's by capillary action (extremely narrow tubes drawing liquid upwards), transpiration (tree breathes out water vapour, dragging the water supply through the tree), or osmosis (magical levelling out of sugar concentrations from cell to cell), or a mixture of all three, it's the kind of thing that if I think about it for too long I need a bit of a sit-down and a cup of tea with extra sugar.

I straighten out the crick in my neck. The screech of a jay rips through the air from the wood opposite. The jays have been out in force recently, the coming of autumn prompting a cascade of acorns. The oak has scattered them far and wide – up to ninety thousand of them in a season. Most are eaten; lots of others rot above ground. Still others are buried by squirrels and jays. Their ability to remember the location of these acorns and come back to them later is enough to impress someone like me, who has difficulty remembering where he put the car keys. But even squirrels and jays sometimes forget, and so an oak tree is born. If there were wild boars knocking about the place they would play their part as well, trampling them into the ground as they ran amok among the green spaces of south London.

I take a step back and look at the oak. In its lifetime it will host millions of animals from maybe a hundred or so species. It was here long before me; it will be here long after me.

There's nothing like nature to make you feel inadequate.

I leave the tree behind and enter the wood, the continued screeching of the jay luring me in. It's not a large wood, but if you stand in the right place it's possible to believe you're not in London at all.

It's here I come in spring for the occasional pre-dawn immersion in birdsong; here I stop off at the end of each walk before re-emerging to a world of tarmac and bus stops and discarded Calippo wrappers on the pavement; here, with eyes closed and imagination working overtime, I can think myself in a larger place, the mix of oak and beech and ash* and plane and sycamore extending far beyond its current, fragmentary existence.

A man is walking towards me, a dog close at his heels. Young, stocky, heavily tattooed, roll-up in mouth, baseball cap backwards on head, white Apple earbuds.

And that's just the dog.

My general approach to random greetings is mixed. Some days I'm happy to avoid eye contact, to mind my own business and let them mind theirs; occasionally I'm nauseatingly chipper, fixing my adversary in my gaze from twenty yards away and assaulting them with a cheery 'morning!' that positively demands an appropriate response. But most of the time I'm non-committal, happy to say hello if hello-saying is called for, but equally content not to invade the personal space of strangers.

* The tree whose scientific name, *Fraxinus excelsior*, I can only hear roared in the voice of Brian Blessed.

With Tattoo Guy I don't even get a chance to choose a tactic. He's seen the binoculars loosely held in my left hand.

'Birdwatching?'

I nod. He stops, ready to talk. His dog sniffs around my ankles. I am a dog lover, but six months earlier a dog of exactly this type – small, yappy, irresistibly attracted to middle-aged male flesh – bit through my trousers into my calf, necessitating a five-hour Sunday evening stay in A&E, and I've been wary ever since.* I maintain my sangfroid with what I consider admirable restraint, and brace myself for the inevitable opening gambit, the default words, the ones that should be printed in bold letters in the information leaflet when you buy a pair of binoculars: 'WARNING: CARRYING THESE IN PUBLIC WILL INDUCE PASSING STRANGERS TO ASK IF YOU'VE "SEEN ANYTHING INTERESTING?"'

But life is full of pleasant surprises. He dives straight in.

'Tell you what I like. Pigeons. Not the scrawny ones, but those plump ones with the white collar.'

'Wood pigeons.'

'That's right. When you hear their wings. You know that? Great sound. Loads of them about, aren't there?'

'Th—'

'And those parakeets, rowdy buggers, but aren't they beautiful? That green.'

'I—'

'They're nesting just over there.' He points across to the other side of the wood. Right on cue, a parakeet jinks across our sightline

* The dog's owner said, 'Oh he never does that', an assertion I found it easy to contradict on the grounds that the dog had literally just done it.

in the distance, a bright green dart in the morning shade. A shrill squawk and it's up into the canopy. There's a short silence. 'And the crows, I like them.'

'Me too.' A carrion crow chooses that moment to hop up onto the temporary fencing a few yards away. We look at it for a while.

'I always look out for the birds when I'm taking him out.' He gestures towards the dog, which is sitting patiently in a 'not about to bite anyone in the leg' kind of way. 'It's just really nice, isn't it?'

I agree. It is really nice. And this everyday, inconsequential encounter cheers me more than I can possibly let on, not least because he's barely let me get a word in edgeways. The people I see in this area, whether strolling through this bit of woodland, taking the dog for a walk round the common, or sitting on a bench by the more formal garden area, don't yield their secrets willingly. I wonder whether they're actively appreciating their surroundings, whether they're there just because a greenish backdrop is more pleasant than a greyish one, or whether they just come there out of habit and because there's a cafe nearby. It's near impossible to tell just from looking at someone what their level of interest is. Unless, like me, they're carrying binoculars. My binoculars are a signifier – I rarely see anyone else carrying a pair in London, and if I do, I'm 99 per cent certain they're a birder – and on this occasion they've enabled an interaction that otherwise wouldn't have happened. I toy with the idea of telling him how grateful I am for his contribution to my day. Because that 'seen anything interesting?' question is one I've always found impossible to answer.

'Ah, well, now you're asking, my friend. Now you are asking. I've seen the first smile on the face of a newborn child; the internal workings of a Wankel rotary engine; morning frost on a pristine lawn; the effect of top spin on a ping-pong ball; the haphazard geometry of a pile of books; sand behaving like a liquid when you pump air through it; Jack Nicholson's face; the rise of the people and the collapse of nations; Tiger Woods' miracle shot on the sixteenth at Augusta that time; a seven-year-old child solving a Rubik's cube in ten seconds; Portland lighthouse rising from the fog as if suspended in mid-air; the mass mourning of Princess Diana; the tears of joy in the eyes of a woman when, after a life of deafness, she hears for the first time; Rembrandt's self-portraits; golden syrup being poured into a bowl; the behaviour of people at bus stops; the languid majesty of David Gower; the miraculous effect of a word of kindness towards a stranger in distress; *The Princess Bride*; the Cruyff turn; a dog on a skateboard; a really good dry-stone wall; a lone butterfly striking out to sea against a deep blue autumn sky; an amplifier that goes up to eleven; the fractal geometry of Romanesco broccoli; the everyday miracle of oh wait hang on you just meant have I seen any interesting birds today, didn't you?'

'Yes.'

'Oh, right. No, nothing, I'm afraid. Sorry.'

One day I might say it. But for now I just take my leave with a smile and we go our separate ways. In another universe, the encounter might have consisted of no more than an exchange of grudging London nods. But I'm glad it didn't. This short, inconsequential meeting, a shared point of contact between two strangers, gives me cheer. And I realise that four years earlier, when I was shy about my enthusiasm and kept my binoculars hidden until the last possible moment, we would have passed each other without acknowledgement.

*

7 May 2019.

I've been twitchy all week, hoping they might somehow come back a couple of days early, but knowing, from previous years, that they're not due yet. There's a nervousness, too. What if they don't make it? Numbers countrywide are down. Something to do with weather systems in southern Europe, maybe; something to do with a general decline in recent years; something to do with ecological collapse.

But ours will make it. They have to.

They do.

Some years, they announce themselves with a scream and a fly-by and an exuberance entirely out of keeping with the distance they've travelled. If I'd just flown in from Africa, I'd barely have the energy to dump my bags in the hall before flummocking on the sofa for a week. But then I have the luxury of a freezer full of food and, should the need arise, a takeaway menu. The struggle for survival is rather more claw-to-beak for birds.

This year their arrival is more discreet, and I see them only because I'm looking for them – and I mean *really* looking, with the kind of attention that nearly gets me run over by an incredulous, gesticulating cyclist as I cross the road at the bottom of the hill. And there they are, wriggling punctuation marks high over the house.

Why swifts? Why not swallows, house martins, sand martins, nightingales, redstarts, chiffchaffs, willow warblers, whitethroats, blackcaps and any of the other birds that make a similar, equally hazardous journey every year?

Well, they're all great too, obviously.

But swifts have something about them. That sickle shape, their astonishing speed, the brevity of their stay in this country* – all contributing to my blind, unrequited love for them, my desire to stand outside the back door looking gormlessly upwards, drinking in their every appearance. Perhaps it's their mastery of the air, the sheer audacity of their aerobatics, the impression they give, with their high-pitched squeals, that they're having enormous fun.† They are, more even than kestrels and albatrosses and hummingbirds, the embodiment of flight. They fly the way I'd like to, if I had the ability – fast and free and screaming with the thrill of it all.‡ Swifts abhor the ground to the extent that if they land on it they need a helping hand to get up again, and when not sitting on eggs they spend the majority of their lives in the air – not like those lightweights, the swallows, with all their perching and whatnot.

I started noting the date of the swifts' return a few years ago, and it proved the beginning of what for want of a better word I call a 'nature diary'. It's a haphazard, woefully inconsistent mishmash of notes and photos and question marks and incomplete lists that would horrify any true naturalist, but it is still, in its own way, a nature diary.

From it I know that on 3 October 2015 I was excited by a blackbird singing from the top of next door's tree, that on 14 April 2018 I saw the first orange tip butterfly of the year in West Norwood

* Late April to early August, and then they're gone. Sand martins are here from March to October.

† I know it's not scientific, but we cannot discount the possibility. Who knows? I certainly don't.

‡ In my head, that is. The reality would be that I'd be slow and lumbering at best, and would have to stop regularly for a breather and to make 'oof' sounds and complain about the pain in my wing.

cemetery, and that every year for the last six years the swifts have arrived on either 6 or 7 May. They're nothing if not consistent.

From it I also know that I have a tendency to forget about my nature diary for about three months every July. 'Twas ever thus.

Sometimes my entries comprise nothing more than a list of birds seen – a dull, efficient way of keeping records, but sadly lacking in the 'fond memories' department; sometimes the shortest of notes evokes a particular image, and I can remember the occasion clearly: 'blue tit on bathroom ledge', or, even more memorably, 'magpie in kitchen'; and when I read the words 'buggy mummies – swans', I remember the warm day in March 2017 when, while admiring two mute swans preparing their pitch for the season's nest in Crystal Palace Park, I heard a young mother say to her friend, 'Oh look, the swans are making a nest. I suppose they'll be laying eggs soon . . . Do swans lay eggs? I should know this.'

These records play their part in the gradual accrual of familiarity with my patch. But if my efforts to get to know it have been earnest and sincere, and if I am beginning to recognise subtleties, changes and patterns, can tell you which birds are most likely to be found doing what and where and when, and if I have started to identify butterflies, trees, flowers and even the occasional bee, then these efforts can only be considered paltry when placed alongside the man whose observations of his patch led to him being known as 'the father of ecology'.

Gilbert White liked a swift. In the second part of *The Natural History of Selborne* – the sequence of letters to Thomas Pennant and Daines Barrington published near the end of his life in 1789 – there are four letters about the 'hirundines', the family then thought to include not just swallows and martins but swifts as well.* The letter about 'the swift or black-martin' is a fine example of White's powers of detailed observation, or as he put it, 'watching narrowly'.

Here he is on swifts:

> It is a most alert bird, rising very early, and retiring to roost very late; and is on the wing in the height of summer at least sixteen hours. In the longest days it does not withdraw to rest till a quarter before nine in the evening, being the latest of all day birds. Just before they retire whole groups of them assemble high in the air, and squeak, and shoot about with wonderful rapidity.

I like that. I like the observation itself, made without optical aids of any kind, but with an understanding not just of what to look for, but of how to look for it; I like everything the two simple words 'and squeak' bring to the sentence; but most of all I like the pleasure redolent in 'shoot about with wonderful rapidity'. These monographs were published in the journal of the Royal Society. They were scientific by inclination, but White's relish of the subject at hand bubbles up from time to time, adding an extra dimension to his descriptions of life on his local patch. *The Natural History of Selborne* combines keen observation with meticulous but not exhausting description,

* White notes in the letter the suggestion from 'a discerning naturalist' that the disposition of the swift's toes (all four facing forward) might mean that 'this species might constitute a *genus per se*'. The discerning naturalist was Giovanni Antonio Scopoli, and he was right.

the ability to see details others had missed, and the willingness to devote many years[*] to the apparently commonplace – and its presentation as a series of letters[†] gives it a personal touch where a drier record might have made for dull reading.

Very little escaped White's attention, and his interests encompassed all manner of species.[‡] He discusses the drinking habits of pigeons, the hibernation of hedgehogs, the tonality of owl hoots, the mysteries of migration and much more besides.

White wasn't the only one recording the ways of nature. He corresponded with other naturalists, comparing notes, learning from them, exchanging ideas. But he mostly stayed at home. Not for him the grand tour, travelling the world recording the exotic and outlandish. Despite suffering from coach sickness he did travel around the country, but his real interest was the area around the village of his birth, Selborne in Hampshire, where he was the curate for forty years. There he observed and recorded, beginning with his *Garden Kalendar* in 1751, and gradually expanding it to include meticulous weather records and eventually a general nature journal.

The breadth of his records, the perception of his observations, and the style with which he recorded them, set him apart. But most of all, it's a record of a single place, compiled over many years by someone who knew how to look, look once more, and look yet again.

[*] He prefaced the above excerpt with 'my assertion is the result of many years' exact observation', and similar phrases are dotted throughout the book, as if keen to establish his credentials in the face of disbelief.

[†] The book is, in part, a sort of fiction. While most of the letters are part of a real correspondence, the first nine to Pennant were written specially for the book by way of scene-setting, while others later on were also never sent.

[‡] With the exception, strangely, of butterflies, which are completely absent. So I'm one up on him there, at least.

It would be rude not to pay him a little homage.

The swifts are long gone when I visit Selborne. We're deep in the heart of autumn, with all that entails: the colours, the leaf carpet, and very much the mud. While the allure of White's house is strong, I've deliberately arrived just after dawn so I can spend as much time exploring the village and surrounding area as possible. Eco-tourist I might be, but I'm keen to put myself out just a bit.

My reward is instant. It's a stunning morning, the kind of dawn that gives autumn a good name. Honestly, there are times when nature seems to be taking the piss.

'Here you go – have something of unfathomable beauty. Here's another. And another. Careful not to faint.'

And so it is that morning. The vista at the top of the Zigzag – the path that White and his brother cut into the steep hill behind the house in the 1750s – would be enough on its own. It's one of those 'looking across the rolling countryside' ones that England does so well.* The balance of fields and woods and white cottages dotted about the place is just right, stretching away to the horizon without the interruption of pylons or power stations or out-of-town shopping centres, and enhanced by the kind of soft, misty, early-morning cloud that hasn't yet decided whether it's going to clear up or close in. It's the kind of scene that would fetch at least a hundred grand at auction if it had been painted by a seventeenth-century Dutchman.†

It gets even better as I make my way along the top towards Selborne Common. The light frost on the grass is one thing; the

* Other countries might excel at them, too, but this one feels particularly English. Please don't write in.

† Who would no doubt have enhanced it with a cow or two.

sun slanting at a low angle through the gaps in the trees, highlighting the russets and yellows and golds and reds, is another; and the halo of light just beyond the bushes, beckoning me to receive a message from God to the accompaniment of a heavenly choir, is one of those One More Things designed to make me crack and admit my helplessness in the face of excessive natural beauty.

And as if that weren't enough, a marsh tit – unknown in my area, so a sighting of some significance for this poor town boy on day release in the countryside – calls to get my attention, pops out in front of me, holds the pose for long enough that I can drink it in, then disappears into the undergrowth. Overcome by picturesqueness, I imagine fondly that this marsh tit is directly descended from the one observed by White, which 'begins to make two quaint notes, like the whetting of a saw'. And maybe it is. It's not lost on me that this description, a reliable touchstone in White's day, would barely be recognisable to anyone nowadays. The nearest equivalent I can think of would be the sound made by the tiniest of laser guns in a cheap video game on your phone, which I'll admit doesn't quite have the same evocative ring.

What with the marsh tit and the frost and the sun and the everything, I near as dammit start a jaunty whistle. It's that kind of morning. All it needs is a dog.

Oh look, here's a dog.

I like dogs. I grew up with one – the sweetest, stupidest, loveliest dog a child could ever hope to grow up with* – and while I haven't

* He was a cross between a border collie and a Norwegian elkhound, and displayed vestigial instincts of both those fine breeds. He could most often be found chasing rabbits round the garden – chasing, but never catching.

succumbed to the joys of dog ownership as an adult, I'm always happy to stop and spend a minute or two communing with one.* Truth be told, they're often better company than their owners. Dog ownership has many benefits, not least of which is the opportunity – obligation, really – for daily exercise. And as long as they're not rampaging through the undergrowth, disturbing nesting sites in the breeding season,† they're a welcome addition to any walk.

This one's a labrador – a good one, too. It's one of those dogs who elicits an involuntary smile and a 'hello there'; the kind of dog who might look at you with big brown eyes and make you, by some mysterious process, reach for the lead; the kind of dog you want to squat down beside and nuzzle; the kind of dog, in short, who might be seen as the purest embodiment of 'good boy'. It responds to my welcoming head-ruffle with a pair of muddy paws on the front of my jumper.

Good boy.

I only have myself to blame. No dog can be expected to resist a squatting human with their arms held open in a welcoming embrace. I brought it on myself. In recognition of this, the owner, following a few yards behind, doesn't bother apologising, instead inclining his head as if in assessment of my behaviour. The old adage about dogs resembling their owners doesn't hold here. The dog is warm and friendly; its owner reserved and tweedy.

Nevertheless, still breathless from the combination of the obscene beauty of the morning and the briskness of my walk, I ride

* Except small yappy-type ones that bite me on the calf (see above). They, and their appallingly unapologetic owners, can fuck off.

† The dogs, not the owners, although you can never be too sure.

roughshod over his reserve. Perhaps I'm a bit over the top. Perhaps 'absolutely stunning' is hyperbole. Whatever the reason, he's not biting. There's a feeling of 'steady on, old chap – wouldn't want to get carried away' about him, and now, before it's even started, our conversation is over. All I get in reciprocation of my enthusiasm is a non-committal 'hmm' through pursed lips, and a 'come on, boy', which I presume isn't aimed at me.

I have a sudden hankering for Tattoo Guy, Apple earbuds and baseball cap and all. We might discuss the bracken, fronds curling and turning to gold in the mid-autumn sun. Or perhaps the little flurry of redwings that choose that moment to *tseep* their way overhead. Or any one of a hundred things that have made the outing so worthwhile. But no. 'Hmm' and 'come on, boy' put paid to that.

And now, as if to punish me for my babbling enthusiasm, the weather turns.

The indecisive clouds have made up their minds. Thick mist it is, with a corresponding drop in temperature. And the early morning briskness I felt when buoyed by the slanty-sunny-misty-frosty idyll at the beginning of the walk dissipates fast, a process aided by a slight turn I give my ankle tripping over a tree root. The ground is boggy, my feet are wet, and I have a hurty ankle. The combination of deserted autumn woods, an empty path snaking round the corner ahead of me, and corvids circling above the misty treetops, *graw*ing and *chack*ing as if they've smelled blood, give me the feeling I'm only a wobbly handheld camera away from a starring role as 'Middle-Aged Victim' in a low-budget horror movie.

Seized by a strong sense of melancholy, and a sudden distaste for all things outside, I turn back, descend the Zigzag – fourteen zigs, thirteen zags – and walk along the road to the warm haven of Gilbert White's house.

Someone has erected a plaque on the wall of the house opposite Gilbert White's – a comical and sly rejoinder to the hagiography of those whose lives were spent in apparently relentless achievement: 'SULLIVAN BLACK, 1720–1793, LIBERTINE, OPIUM-EATER, DRUNKARD, DUELLIST, GAMBLER AND WASTREL, LIVED AND DIED HERE'.

For a moment I think it's dedicated to a real person. Then I make the connection between Gilbert and Sullivan and White and Black, realise that the dates are the same as the great ecologist's, and give a little smile and a nod of acknowledgement to the person who thought it up and went to the trouble of installing it, and it serves as a reminder that many lives are lived in a way that won't be remembered, and that there's nothing wrong with that.

I visit the house. It does all the things it needs to, and more. I read about White's life, his prodigious fruit and vegetable growing, with particular emphasis on cucumbers and melons; I read about the pests – cockroaches and house flies – that infested his kitchen; I read about his desire to emulate in a small way the grand gardens of places such as Blenheim and Stowe, with the building of a ha-ha and a wooden cut-out statue of Hercules.* I'm reminded of his appreciation of the earthworm – 'though in appearance a small and despicable link in the chain of nature, yet, if lost, would make a lamentable chasm' – and think of Charles Darwin, who, having read *The Natural History of Selborne*, wondered 'why every gentleman did

* He couldn't afford a stone one.

not become an ornithologist'. I amuse myself for a while with the interactive test on birdsong, and am relieved to get them all right.

And then I visit the rest of the house, the part devoted to Captain Lawrence Oates and his uncle Frank, and I read about the expedition to the South Pole on which Lawrence Oates and all his colleagues lost their lives, and I think about human bravery – inevitable when contemplating the Scott expedition – but also about the urge to explore, and how it manifests itself in different ways in different people.

And having done all these things, I realise that while this place is wonderful and fascinating and steeped in history, and has offered me an experience not available on my own patch, I will always be a tourist here. And I have a sudden urge for the familiar.

So I drive home and walk around my patch once more, with its view of urban sprawl and its feral pigeons and its Perfectly Normal Tree. In deference to Gilbert White I watch it all as narrowly as I can. There are bees buzzing around flowering ivy, and I lean in to look closely. There are at least three different species, and I make a note to look them all up on my bee chart when I get home, and if the person who walks past at that moment, talking on the phone about interest rates and commission, looks at me a bit oddly as I do all this, then that, frankly, is their business.

4

IN THE PICTURE

In which our author really looks at a heron, faces the reality of 23,000 photographs, drops in on Thos Bewick, and drinks from the abundant keg of avian glory

I have a blank piece of paper, and I have a pencil. I can do what I like with them. I could write a poem, draw a musical stave and compose a ditty, or I could write out from memory Lewis Carroll's 'Jabberwocky', the only poem I learned as a child to survive the intervening forty-five years intact.

I choose to do none of those things. Instead I draw a tufted duck.* Or at least something you could give someone and they might say, 'Is it . . . a duck . . .?'

My credentials as an artist are thin. I'd go further and say they're non-existent. Art has been lodged in my head for decades

* Easier, on the whole, than drawing a slithy tove or a mome rath.

as 'something I can't do', and I've never seen any reason to prove myself wrong. Why would I do that, when the results would no doubt induce queasiness in anyone unfortunate enough to catch a glimpse of my efforts?

My speciality, aged nine, was what I called 'one of my landscapes'. My landscapes consisted of one line gently curving across the page from the left (a hill), intersecting with another line gently curving across the page from the right (another hill). Sometimes there would be a car on one of the hills, sometimes not. According to whim I might also add a bird or two, and a sun (or two).

I never progressed from my landscapes, nor attempted to, and so came to the inevitable conclusion that I 'couldn't do art'.

It is often the case that we are interested in that which we cannot do. My fascination with the works of the great artists is accordingly tinged with an awestruck reverence. This reverence differs from the respect I afford great composers and writers, whose skillsets and tricks I am more or less familiar with. Art, to me, is a form of sorcery.

But sometimes you get the urge to dabble in sorcery.

I blame birding. A couple of years spent practising this noble hobby brought with it a sketchy understanding of its associated skills, known to the experts as 'fieldcraft'. Some improved simply by dint of doing them often. My ability to identify a blue tit from a fleeting glimpse of its disappearing arse, once a pipe dream, is now a source of pride; I can now bring my binoculars to my eyes and focus on the place I need to look at without whacking myself in the nose; and if pushed I can give a quick description of where in the featureless landscape my birding companion should direct their gaze. But if these skills are in my grasp, or at least near to it, I have resolutely resisted developing one of the most important: sketching. I have a camera. If I want to record a bird for later identification, I use that – quicker, easier, and no faffing around with sketchbooks and pencils and having to be good at it.

But then I see other people's sketches, posted online, and the prospect seems somehow alluring. Just a few lines to give the overall impression of the bird – body and head shape, tail length, distinguishing features like eye stripes and wing bars and such like. Wouldn't that be a good thing to be able to do?

It would.

Not trusting a live bird to sit still long enough for my purpose, I go to my photo collection. Soon enough a suitable candidate presents itself. A heron,* standing in beaky isolation against a backdrop of reeds. Should be simple enough. Herons are monochrome birds, well suited to pencil sketching. And they have a distinctive shape, recognisable to even the most casual park visitor. Pointy dagger beak, round eye, scraggly shawl. That should do it.

It doesn't.

The bill is unequivocally the wrong shape, the head is more like a donkey's, and is somehow hanging from the body in a way that makes it look as if the bird is halfway through being beheaded. And quite what a heron is doing wearing a toupee is anyone's guess. It looks like a heron drawn by someone who has heard a five-second description of a heron, in a language they don't speak, by someone

* Ironically enough one of the few birds that is likely to stand still for a long time while I fuss around with sketchbook, pencil, and especially eraser. But why brave the freezing wilds of Brockwell Park when I can sketch in the comfort of my own sofa?

who once met a guy who heard about a heron that visited their grandmother's home town fifty years ago.

I try to fix it, changing a line here, adding one there, rubbing out the toupee and drawing a crest.

Now it just looks like a heron that has been drawn by a heron.

Drawing is every bit as hard as I thought.

I look again at the photograph. It's a lovely one, the bird's profile strong against a blurry background, and with nothing intervening to complicate matters. Why is it so hard?

I abandon the drawing and try again.

My second effort looks like the work of a bright six-year-old.

I call that progress of sorts and, undaunted, have one more go. And it's only now that I realise the basic error of my ways.

I have never looked at a heron. Not properly.

I've looked at them many times, both with the naked eye and through binoculars. I recognise the shape instantly, whether it's standing with infinite patience in the local park, flying overhead with those slow, lapping wingbeats and distinctive bulgy neck, head back between its shoulder blades, or, as on one memorable occasion, landing on our shed roof, scrabbling against the unexpected camber, then flying away with an aggrieved *grawk*.

I know what a heron looks like.

But at the same time, I realise, I don't know what a heron looks like. There's a difference between recognising something and knowing how it's put together. It's the difference between being able to order *paella y dos cervezas, por favor,* and actually holding a conversation in Spanish.

I spend a few minutes examining the photograph, taking in the precise angle of the tip of the bill, the rate at which it thickens towards the head, the precise configuration of its base, how that relates to the elongated nostril on the top. The eye – near the front of the head, level with the top half of the bill – has a thin black ring surrounding it, and then there's a thick black stripe sweeping back, dominating the top of the head, thinning out a bit, and topped with a few wispy streaks of feather which form a perky little crest.

This realisation is one thing. Making my hand reproduce it is quite another.

I make a tentative line – the top of the bill. It wobbles. I try again, willing myself to be more confident. Now it's too long, too heavy.

Third time lucky.

Bingo.

And now the bottom of the bill, being careful – but not too careful – to get the angle right.

I build the basic outline of the bird, referring back to the photograph time and time again to check my progress, and eventually I lay down the pencil. It's as good as I can make it. It's finished.

Maybe just another line here?

No. Leave it be. You'll ruin it.

But there's—

No.

It's the easiest mistake to make. Fiddling for fiddling's sake. Thinking that the next little amendment will be the thing that transforms it, when in fact you might have been better off getting it right in the first place.

I look at my heron. At the very least, it resembles something its mother might recognise, so I'm taking that as a win.

It's not just the sense of achievement that gives me a boost.

There's something more. I realise that, while I was drawing this heron, my mind was occupied with one thing only: drawing the heron. Whether you call it mindfulness or meditation or just concentration, that focus – the single-minded devotion of one's energies to completing a task – feels like a healthy thing. Throw in the strangely relaxing physical sensation of making the marks – the smooth sweep of drawing a curved line, the staccato tapping when doing the speckly bits, the obscure pleasure gained from filling in the eye with little circular movements until there remains the tiniest speck of white, the life-giving gleam of it – and you have pure therapy. And I've ignored it for nearly five decades.

What a pillock.

An old room in an old house. I'm alone. Alone, that is, apart from seven black and white heads, each one much larger than life in a way that could easily distress me if I thought about it too much. They're papier-mâché models of seven British birds: heron, cuckoo, roseate tern, tree sparrow, blackbird, dotterel and great black-backed gull.

There's a circle of chairs in the middle of the room. I can take my pick. I choose to sit under the baleful gaze of the heron.

Dotted around the room are speakers, voices talking.

'... I feel powerful. It's a food chain ...'

A human, pretending to be a bird.

'... they're all killing something ...'

The words, if I'm not mistaken, of a great black-backed gull.

'... we just happen to be killing something people like ... like puffins.'

It has a point, you know.

My visit to Thomas Bewick's house, Cherryburn, has coincided with an installation, and it's very much my kind of thing. The artist Marcus Coates has gathered leading figures in the world of nature in a recording studio and asked them to talk about their lives, the only trick being that they have to do it in the character of a bird of their choice.

'. . . the birds that are thriving are the ones that have found niches as part of the human world . . .'

A few people look in at the door. They don't stay, possibly put off by the massive bird heads dominating the room. Outside, a wood pigeon sings its lilting, syncopated song. 'Ho-HOO-hoo, ho-hoo. Ho-HOO-hoo, ho-hoo.'

'Never trust a bird with a penis.'

Helen Macdonald, author of *H is for Hawk*, having a whale of a time being a cuckoo.

'I have an unjustified bad reputation. Humans hate cuckoos because we reflect their own anxieties about whether their children are theirs.'

She's really getting into her role. This is high-level anthropomorphism – frowned on by scientists, but almost inevitable nonetheless. We can't know what goes through the head of any animal, and any attempt to do so can be made only from a human perspective. And it's enjoyable listening to knowledgeable people trying.

'The migrants are intruding.'

Geoff Sample, naturalist and wildlife sounds expert. Also, for now, a blackbird.

'They come over here, from Scandinavia, invade our territory, compete with us for resources . . . young blackbirds muscling in on my patch. The old blackbird culture's not there any more. I'm getting old . . . might move out to the country . . .'

The discussion is played on a loop. I listen to it all twice, sitting quietly on the hard chair, drinking it all in. On the one hand, it's a playful exercise, but it also encourages us to look at things from a different angle, in a different way. It's one thing looking at something closely so you can draw it; quite another trying to get into its head.

'. . . in the end, you learn from whoever's around . . .'

The recording reaches the beginning again. It's dark in the room, warm and sunny outside. I go out to the yard and sit, looking over the Tyne valley across the Northumbrian countryside. Cobbled yard, low brick wall, trees to one side, rolling hills stretching to clear blue sky. I can see why young Thomas bunked off school.

A small tortoiseshell butterfly lands next to me. Colourful, delicate. Lovely. If it were a moth, brown and drab, most people would dismiss it. Our view of nature is relentlessly coloured by sentiment, our insistence on imposing human values on non-human things. Colours and patterns are interesting; plainness isn't. Gulls are bastards, robins sweet, otters cute, snakes slimy.

But if you look at behaviour, not image, none of them are any of those things. And they're all here, alongside us, for one reason only: they have survived.

The small tortoiseshell basks on the wall next to me, wings spread. I open my sketchbook and do a quick, inaccurate sketch, deficient in almost every respect. It flies away as if disgusted.

A blackbird hops up onto the wall, levers its tail slowly upwards. I have my sketchbook open. I also have a camera. As I switch it on and wait the few seconds for the zoom to reset, the blackbird cocks me a look and flies away.

Cheeky buggers, blackbirds.

In 1789, at the same time as Gilbert White was publishing one wildlife masterpiece, Thomas Bewick was putting the finishing touches to another.

It was a time of burgeoning interest in natural history. Explorers were coming back from far-off lands with descriptions of exotic beasts;* the great and the good assembled collections of one thing after another – flowers, shells and, where budget and space allowed, animals; and scientists were busy nailing down the vexed question of what exactly was what. Over a hundred years earlier, the great naturalist John Ray had classified over two hundred British birds; Georges-Louis Leclerc, Comte de Buffon's monumental 36-volume *Histoire Naturelle, générale et particulière,* with illustrations by Jacques de Sève and François-Nicolas Martinet, appeared piecemeal through the second half of the eighteenth century; and the contribution of Swedish naturalist Carl Linnaeus, whose binomial classification system is still in use today, took the science to another level.

Bewick sensed that a reliable, accurate and exhaustive survey of animals would find an eager audience, and since 1781 had been planning such a book. He had already produced thumbnail-sized engravings for *A New Lottery Book of Birds and Beasts* in 1771, but the plan for *A General History of Quadrupeds* was far more ambitious.

* Not always accurate, as Joseph Banks's description of a kangaroo – 'as large as a grey hound, of a mouse colour, and very swift' – illustrates.

He worked with Ralph Beilby, the Newcastle engraver to whom he'd been apprenticed in 1767, and in whose workshop he had learned the intricacies of wood engraving. Beilby wrote the texts describing the animals, although these were also liberally peppered with memories from Bewick's own childhood. They drew on the work of Ray, Buffon and Linnaeus, as well as Thomas Pennant, author of *British Zoology* and one of Gilbert White's correspondents in *The Natural History of Selborne*. The clash of the various classification systems caused them endless headaches when it came to working out the animals' order of appearance in the book, a dilemma they never truly resolved. But if the resulting book, for all its many qualities, was an energetic muddle, it proved a popular one.

It's easy to see why. *A General History of Quadrupeds* mixed the familiar – horses, cows, sheep – with the outlandish – zebra, dromedary and the wonderful cameleopard, or what we now know as a giraffe. Beilby's descriptions of the animals are successful enough, but it's the illustrations that give the book life and popular appeal: more than two hundred wood engravings of animals, plus another hundred or so 'tale-pieces' – vignettes of country life, usually telling a little story – at the end of chapters.

Unsurprisingly, the most successful illustrations are of the animals Bewick knew best. A 'cart horse' exudes lazy charm, rough of coat, heavy of limb, its somnolent eyes half-closed, contrasting with the wide-eyed and well-groomed 'improved cart horse' of the following plate. A common goat, with straggly hair and beard, has elegantly curved horns and an indefinably goaty look in its eye. And he somehow captures the blend of vulnerability and strength in a recumbent fallow deer, its palmate antlers delicately perched on its head.

Bewick's desire for anatomical accuracy was somewhat undermined by an inconvenient truth: he had never seen many of the animals he was portraying. Relying on other people for drawings and descriptions naturally compromised the accuracy of some of the illustrations, so if the folds of skin on his rendition of the rhinoceros are somewhat exaggerated, giving it the impression of wearing a coat several sizes too large, this is both understandable and forgivable. It is in any case a striking and memorable image, as is the magnificently ungainly dromedary – a lumpy and shambolic animal of the kind you'd like to get to know.

And then there were the 'tale-pieces', which exuded a lively wit, and packed a lot into a small space. In one of them, two people try to get a mule going – the rider by waving his hat, the accompanying pedestrian by flaying it with a branch. A windmill behind them recalls Don Quixote – the absolute stillness of the mule an accurate rendition of the impotence of humans in the face of animal stubbornness. Another shows a man trying to hold his balance in driving rain on a narrow plank over a stream, his hat blown off and his dog straining at the leash. A third has a man with a bunch of sticks on his back, small dog in front, trees with rooks' nests in the background, the weariness evident in the stoop of the man's back, the dog bounding off ahead of him.

For a reading public newly switched on to the wonders of nature, these illustrations would have opened up a new world of understanding, both of the natural world around them, and of the unfamiliar and exotic. And it was all underpinned by a level of observation that had become a habit since a childhood spent skipping school and roaming the countryside around his home, drawing everything he saw.

Bewick's preferred medium, engraving on close-grained boxwood, was out of fashion when he started his apprenticeship. Recent advances in metal engraving offered cleaner, more elegant lines than the somewhat rough standard prevalent in the woodcuts of the time. But Bewick felt at home with wood, and cut his teeth in Beilby's workshop, learning and perfecting the techniques that would serve him for the rest of his life. It helped that he had an innate ability to see and think in negative – backgrounds were black, the design coming from what was taken away, not what was added. But most of all he saw that you could produce variations of tone by adjusting the depth of incision in the wood, a technique not available in metal engraving. He also developed a deep understanding of how best to print from woodcuts, which meant he could produce work of superior detail and quality. And his manual dexterity and control – as with all things, a mixture of natural ability and practice – allowed him to achieve an unmatched level of fine detail.

The final engraving in *Quadrupeds* is the only one without a name:

> An Amphibious Animal . . . about the size of a small cat . . . its bill is very similar to that of a duck . . . its eyes are very small, it has four short legs . . . the fore legs are shorter than those of the hind, and their webs spread considerably beyond the claws . . . the hind legs are also webbed.

Given that the waterways of Northumberland weren't exactly overflowing with them, the accompanying illustration of a duck-billed platypus isn't half bad.

Quadrupeds was just the beginning. Bewick's biggest project, the one for which he became known, and the reason his name has been given to two bird species,* was *A History of British Birds,* published in two volumes over eight years around the turn of the eighteenth century. It was a mammoth undertaking, not made easier by the influx of work – ranging from coats of arms for public offices to bar bills for local taverns – that resulted from the popularity of *Quadrupeds.*

British Birds builds on *Quadrupeds* in every way. The engravings are more uniform in quality, less prone to inaccuracy. The descriptions† are more detailed and evocative. And the 'tale-pieces' have developed, some of them packing astonishing amounts of detail into tiny spaces, and many offering trenchant social commentary in the scenes they depict, giving us a picture of rural life of the time. In a way, for all the brilliance and importance of the animal illustrations, it's these vignettes that tell us more about the man and his times.

Bewick's mastery of wood-engraving techniques allowed him to show the birds' plumages in extraordinary detail, somehow rendering the subtleties and nuances of soft materials in a hard medium.

Some of the birds sit nicely, as if for a portrait. Not so the cuckoo. It's caught in mid 'cuck', mouth open, tongue half out, chest thrust forward, long tail reaching for the sky at a defiant angle. The upper parts, slate grey in real life, are composed of innumerable fine notchings, sweeping back from the head along the back, subtle variations in length and intensity portraying the plumage's infinite nuances, delineating the join between body and shoulders, the layering of the

* Bewick's swan (*Cygnus columbianus bewickii*) and Bewick's wren (*Thryomanes bewickii*).

† Written by Beilby for the first volume, but by Bewick alone, following a falling out with his long-time partner, for the second.

wing feathers. And if the monochrome illustration can't capture the yellow around the eye, it compensates by somehow showing us its livid gleam, portrayed, I now see, not by showing it as a circle, but as a crescent moon, surrounded by the finest filament for an eye ring.

What he has captured, I realise as I examine this bird in minute detail, is the cuckoo's cuckooness.

I think of him sitting at his table, working with concentration on these fine details, matching the tool for the task, moulding the wood with craft and patience, varying the depth of his incisions to give the nuances of tone and detail that set his work apart. A serious, jocund, perceptive man, a rare and original artist whose work introduced thousands of people to the world of nature.

I look at his cuckoo, take a piece of paper and a pencil, and try again.

The seven birds featured in *Conference for the Birds* represent a decent cross-section of British species, from ubiquitous to scarce. If you're not going to struggle to see a heron or a blackbird, the same can't be said of the tree sparrow – population down more than 90 per cent in the last forty years – or the roseate tern – now restricted to a few coastal breeding sites in Britain. But if you had to pick one of them as an example both of a famous British bird and one in steep decline, a lot of people would choose the cuckoo.

You don't often get to see cuckoos – they're relatively secretive birds. Nowadays, you don't often get to hear them either.

But sometimes you just get lucky.

There is a good place for them. I try to get there at least once every year. You can never be sure – they might not have arrived, or they might be out and about that day – but the RSPB reserve at Northward Hill in Kent is a reliable cuckoo haunt.

My desire to seek them out is, I suspect, fuelled by a desperation to cling on to the past, to recapture a sliver of something that in my childhood was as natural as breathing. The cuckoo's two-note call was a summer constant, just part of the soundtrack.

I go there for the nightingales too. Another lost sound. And it's the nightingales I hear first.

I stand where the path divides. There is a terrific greenness all around, bushes and shrubs and trees all throbbing with life, possibility and, frankly, sex.

In the background, a chiffchaff – monotonous, repetitive, but somehow uplifting. Further away, a blackbird, the complexity of its song putting the chiffchaff to shame. All around, a selection of other hopefuls, their vocal fecundity driven not by a desire to please the human ear – although this is a bonus from our point of view – but by evolutionary imperative. I stop and feast my ears.

And then it starts, not far away, dominating the soundscape as if amplified by the bushes. The richness of the nightingale's song has been celebrated since time immemorial, and it's one of those things you have to experience live. Recordings, though useful as a guide to recognition, simply don't do justice to the depth and richness of the sound. And here there are two of them in competition.

I drink from the abundant keg of avian glory until I've had my fill, and then I move on, walking slowly, as the idyllic nature of the scene demands.

A brouhaha overhead. An aggrieved chattering, a kind of strangled squawk. Birds unknown, bothering each other high in the canopy.

I'm better than I used to be at finding these things, at hearing something, locating it, synchronising sound with sight. But still it eludes me. The canopy is too thick.

The brouhaha intensifies. It's now verging on a ruckus, possibly even a stramash. And with the stramashery comes visibility. Branches sway, leaves flurry, and the two birds tumble briefly into view, the larger bird landing, to my intense satisfaction, on a branch to which I have a direct sightline.

It's a cuckoo, and it's mighty pissed off.

I can see the barring on its front, the livid eye, the head turning this way and that to fend off the bombardment of fury from the smaller bird, which doesn't stay still long enough for me to get a handle on its identification. And the sounds it's making are at such a pitch of incoherence that they're unlikely to appear in any lexicon of bird sounds except under the heading 'Generic Small Bird Fury, Incandescent'.

An image pops into my head of King Kong at the top of the Empire State Building, swatting away the swarming planes.

The battle between the two birds transfixes me, and then, as soon as it started, it's over. The smaller bird, identified as a whitethroat just as it departs, gives up the ghost, and no sooner has it disappeared than the cuckoo ups sticks and the canopy is restored to a semblance of peace.

And here's the thing.

I had my camera. It would have been the work of a few seconds to bring it to my eyes and fire off a burst. Point, press, hold. Twelve shots a second, maybe five seconds' worth. Point again, press again. Repeat. Upwards of a hundred, two hundred photos through which to sift at my leisure, deleting the unsatisfactory ones until a handful remain. And that handful would be 85 per cent leaf, 10 per cent blur and perhaps, if I got really lucky, 5 per cent aggrieved cuckoo face. Maybe.

But none of them would come close to the memory of it. And it's just possible that by not photographing the scene, by focusing instead on drinking it in, observing every detail – *experiencing* it – my mind holds it in the memory just that bit more firmly. Apart from anything else, if I'd opted to get out the camera my focus would have been on getting a decent photograph rather than the scene itself.

The absence of a bad photo from my library hasn't expunged the incident from my mind. Far from it. And when I sift through my photos, gradually reducing the strain on my hard drive, there are any number of images that seem pointless and one-dimensional compared with the vividness of that cuckoo.

I've taken thousands of photos. Birds, deer, trees, butterflies, lizards, dragonflies, rabbits, bees, spiders, flowers, snails, bracket fungi sprouting from tree stumps, mosaics of autumn leaves, patches of scrubby foliage nudging through the bars of cemetery railings, early-morning sunlight streaming through the gaps in the canopy, fields shrouded in atmospheric mist, bluebells carpeting the woodland floor.

And ducks. Loads of ducks.

I do this mostly because I find these things attractive and would like a record of them. I also do it because it is cheap and easy. Too cheap and too easy. Phone out of pocket, point, click.

At the time of writing I have nearly 23,000 photographs on my computer. If I took the time and trouble to delete the dross, duplicates and inexplicable screenshots, no doubt I could bring that number down by at least 50 per cent. But that's still too many, the equivalent of whole drawers of prints, none of which ever get looked at except when you decide to have a good clear-out and then spend an afternoon sifting through old photos – oh my God, look at our hair! – before putting them all back in the same drawer.*

Why exactly do I have thirty-six bad photos of a black-headed gull? Because I couldn't be bothered to delete them. Do they remind me of the day, place me back in the moment, experiencing what it was like to see that black-headed gull? Do they perhaps give a unique insight into the bird's psychology or behaviour? Do they serve any purpose at all?

Nah. I probably took them because I thought I'd get a few likes on Instagram.

Some of them stick. Looking at a particular photo of two hundred jackdaws and crows spreading across a dusky winter sky, I am taken back to that chill December day in Gloucestershire, can feel the crisp of frosted mud underfoot, the bite of the wind on my gloveless fingers. And a wren, perched on a reed, head thrown back, mouth open, pouring its punchy song up to the sky, places me on the path in the warm spring sun at RSPB Rainham Marshes, a general sense of well-being enhanced by the exuberance of that tiny shouter.

And if some of these photographic success stories serve a purpose, triggering memories or acting as a record of an occasion, there are others that are interesting or worthwhile images in their own right, eliciting a response in other people. It might be because the framing is interesting, the light dramatic, or some other aspect of the photograph that triggers pleasure; or it might just be a happy

* I have those, too, obviously.

accident, a confluence of factors beyond my control, conspiring to make it memorable in some way.

But no matter how striking any of my photographs might be, no matter how nearly perfect, they will always lack the evocative magic of the very first photographic images of nature, taken nearly 180 years ago.

In the 1830s a revolution was afoot. The possibilities afforded by combining specific chemicals and exposing them to light were becoming apparent, and the pioneers of photography were conducting experiments that would eventually lead, albeit indirectly, to that bane of modern life, the selfie. If only one could invent a time machine to go back and nip this dangerous process in the bud.*

In France the pioneers were Louis Daguerre and Hippolyte Bayard; in England, John Herschel and Henry Fox Talbot led the way. And it was Anna Atkins, using a process invented by Herschel, who produced the first photographically illustrated book about anything, anywhere.

Atkins's motivations to produce *Photographs of British Algae: Cyanotype Impressions* were manyfold. She was a keen botanist† – in this she was encouraged from an early age by her father, John George Children, a renowned scientist, secretary of the Royal Society, and notably liberal parent for the time. She was also a talented illustrator, producing 250 engravings for her father's translation of Lamarck's

* I am being a curmudgeon. Pay me no heed.

† In 1839 she became one of the first women admitted to membership of the Botanical Society of London.

Genera of Shells in her early twenties. And she was fascinated by the new discipline of photography.

In particular, she was intrigued by the possibilities of cyanotypes, which her friend John Herschel* invented in 1842. This technique – mix chemicals,† paint them on paper, expose to sunlight, rinse and dry – was quicker, simpler and cheaper than film photography, and while the results were less detailed, they were ideal for Atkins's purposes.‡ William Henry Harvey's otherwise definitive reference volume, *A Manual of the British Algae,* published in 1841, was admirable in many ways, but it had no illustrations. For someone as visually engaged as Atkins this was hugely frustrating, and in cyanotypes she saw the opportunity to produce a companion publication of illustrations to fill the gap.

In recognising the power of the photographic image, its importance in communicating the beauty of nature, and the difference between a human's representation of an object and the real thing, she was ahead of her time. The wood-engraving technique mastered by Thomas Bewick remained the default method of book illustration, but for all the beauty of skilfully rendered artwork, and while a good botanical drawing was still useful as a scientific guide, cyanotypes spoke to people in a different way. Their novelty was one thing – their translucent beauty sealed the deal.

* The son of William Herschel (who discovered Uranus), and a distinguished astronomer in his own right. His aunt was another astronomer, Caroline Herschel – more of her later.

† An 8.1 per cent solution of potassium ferricyanide and a 20 per cent solution of ferric ammonium citrate, chemistry fans.

‡ The cyanotype became the default method of reproducing designs in the architectural trade, known – of course – as 'blueprints'.

Atkins published the first instalment of *Photographs of British Algae: Cyanotype Impressions* in 1843, with further volumes appearing over the next decade. It was painstaking and meticulous work, requiring skill and patience. The specimens were fragile, and had to be handled carefully and quickly, before leaving them exposed to the sun for exactly the right length of time. She printed a limited edition for private distribution – algae, then as now, weren't big business – and today just seventeen copies remain, each one in a different configuration. These aren't things you can just turn up and look at, but you can examine them online.*

The character of the images is enhanced by Atkins's decision to write the text out by hand – a pleasing personal touch. Her handwriting is careful, neat. There is an elegant curl to the loop of her 'g's, a soft rhythm to the listing of the scientific names. *Lichina pygmaea, Polyides rotundus, Porphyra linearis.*

The cyanotypes themselves are striking. Some plants, the less frondy ones, are merely patches of white – blobs, in some cases – against a blue background. But the ones that exert a particular hold are the straggly ones, myriad tiny white lines reaching for the void.

One, *Ectocarpus fasciculatus*, is suspended near the top of the sheet, giving it the impression of floating in the sea, seen from underneath as if you're a diver looking up, the algae silhouetted against the sun. Another, *Gigartina acicularis*, looks like an overhead map of animal migration routes, its paths straying off in various directions, intertwining, crossing each other and gradually fading out until just one faint trace wisps its way towards the top of the page, the lone survivor.

* Thank you, New York Public Library.

You can lose yourself in these images. They offer a pleasing contrast to the sepia tones of other contemporary photographs, and the variety of patterns created by the algae – skeletons in the sea – take on a vivid luminosity against the depth of the Prussian blue background. Their abstract other-worldliness offers respite from the seething abundance of nature images in the modern world, and you quickly become used to it. And their historical importance shouldn't be understated. Those glossy coffee-table books, lavishly illustrated with stunning images of everything from the New York skyline to birds of the world – they all owe a small debt for their existence to Anna Atkins.

There are times, when looking for wildlife, that you know you're going to have a duff day. The ominous calm from the moment you park, the empty lake, the seemingly permanent lull. Today is not one of those days. The moment I see the first flash of blue under my nose, a minute after parking the car, I know I'm in for a good time. It zips across ahead of me and towards the small pond by the boardwalk, a blur of movement and activity in the stillness of the heat haze. I follow its trajectory, find it, head towards it.

A damselfly – along with its dragon cousins, one of the surest harbingers of summer. Their jewel-like appearance and aerobatic mastery combine to exert a rare fascination, although there are admittedly frustrations attached. Their moments of stillness seem perfectly synchronised with my ability to identify them, the simple equation being that they sit or hover still for exactly one second less than the time it takes for me to hazard a guess as to their identity.

This one is, to my relief, one of the few I'm fairly confident of, its metallic blue sheen almost as mesmerisingly attractive as its latticework wings and, indeed, its name. Banded demoiselle. Words that positively invite you to deepen your voice half a step and say them slowly, with an inviting half-smile.

The apparent delicacy of these creatures disguises a robustness that has seen them survive for about 300 million years. Our modern *Odonata* (the genus encompassing both dragonflies and damselflies) might not have the two-foot wingspans of some of their ancestors, but there is still something of the primeval about them. But, most of all, their outstanding aerial abilities make them a source of constant fascination.

The mechanics of their flight enable them to manoeuvre in pretty much any direction, including backwards. Four light, strong wings can produce power disproportionate to their weight, and they can angle their body to enable flight that combines nimbleness with speed in a way almost unmatched in nature. It's taken us thousands of years to develop technology that comes close to what evolved naturally.

This specimen treats me to a short display before zipping across to the other side of the pond, and I continue on my way.

I've come to Thursley Common in Surrey. A 350-hectare area of heathland just off the A3, with terrific biodiversity, it's an enticing venue for a south London wildlife-lover with a yen for something a bit different from the local patch. Where I'm used to blue tits, jackdaws and the occasional squirrel, Thursley offers a broader palette, including the already seen dragonflies and extending to lizards. It's prime lizard weather, so my eyes scan low down ahead of me as I set out along the boardwalk that threads its way across the wetter part of the common, looking by the posts and around the edges for odd shadows, strange disturbances in the regular board-gap-board pattern.

And sure enough, within a minute my vigilance is rewarded. It's average sized for a common lizard, about five inches, sleek of appearance, and, as lizards always are, looking poised for action. It's something to do with the set of their feet.

To take a decent photo I'm either going to have to stand directly above it and look down, emphasising the dramatic silhouette of its body and the barcode pattern on its back against the dull grain of the wood, or I'll have to get down to its level. Not wanting to disturb it, I opt for the latter.

My dodgy knees are already protesting at the thought. This will be worth it, though. And to be absolutely fair, it's not the getting down so much as the getting back up.

I make sure I'm still a few feet away as I lower myself slowly to my knees. The lizard stays there, still, wary.

I lie on my stomach. It's strangely relaxing. And now I inch forward using my elbows until I'm close enough to see its occasional blink. But I'm aware that getting too close will scare it off.

Lizards are, by any standards, fascinating creatures. Like dragonflies, they exude a hint of the prehistoric. More than a hint, in fact. A shedload. If we need a living connection with the history of this planet, lizards tick all the boxes. More so, in fact, than many birds. It's the feel of them, the impression they give of being terribly ancient.

I suspect, though, that many people don't even realise Britain has lizards. They're associated with warm sunshine; Britain is not.

My usual procedure when taking a photograph is to fiddle around with the settings until it feels right. I'm well aware that this is not the prescribed method. I'm also aware of the three pillars of photography – shutter speed, aperture size, ISO – and the effect changing each of them will have on your photograph. I know this, and I understand it. And yet my preferred method is still, and will always be, to fiddle around with the settings until it feels right.

This is why, when I take a halfway decent photograph and people compliment me for it, I flap them away like a spheksophobe dismissing a wasp. Real photographers are rather more organised in their approach to the discipline.

On this occasion, my fiddling proves reasonably effective, and I manage some respectable shots before the disruption arrives.

The lizard feels it first. A distant vibration. And it's off, darting down under the boardwalk, safe refuge from the approaching monster. And now I'm aware of it, can feel and hear it coming towards me. *Thubb – thubb – thubb – thubb.*

It's a jogger. An athletic one, too, springy and lithe and fast, showing all the right signs of exertion without that desperate look you sometimes see in a runner's eyes, that look that says, 'I'm doing this because I think it's good for me, but honestly, kill me now.'

He clumps past, showing no sign of incredulity that there's a middle-aged man lying on the boardwalk. For my part, any incredulity I have is reserved for other people's athleticism, and it's really more of a simmering low-level envy. I prefer walking to running for many reasons, pain and discomfort prominent among them. But while I'm not averse to building up a head of steam on a brisk walk – preferably with the help of a stout stick – I would also be open to the possibilities afforded by more strenuous exercise. I have it on good authority from those who run regularly that the exercise brings with it a heightened appreciation of one's surroundings. But to counterbalance this is the knowledge that nothing beats a really good dawdle. The fast walker or runner might be getting an endorphin hit and a decent cardiovascular workout, but they run the danger of missing out on the goldcrest flitting about in the cypress just there, or that really good bit of hairy lichen.

So I waver between dawdle and stride, never quite summoning up the energy to discover the joys of running.

I haul myself back up and continue, setting my default pace to 'stride with the occasional dawdle should the need arise'.

It arises.

A Dartford warbler plays hide-and-seek behind a clump of gorse; a redstart sings from a nearby copse, pops out to say hello, pops back in again; a willow warbler tirelessly gives its melancholy descending song from a few yards away.

And now a different sound.

Cuck-oo.

My directional hearing isn't what it was, but a few repeats lead me to believe that it's over there. I follow the sound, hoping it won't suddenly decide enough's enough.

It's quite loud now. There's a time for dawdling and a time for striding. Now is the time for striding.

I reach a long row of trees, with a field behind. I can just see enough through them to be aware of a presence. And as I approach, following a rough path through the low scrub and gorse, I catch sight of them.

Photographers. Sitting in a crescent, lenses trained on a spot I can't see through the trees. They're real photographers, too. Male, camouflage gear, telephoto lenses the size of the Blackpool Tower.

I feel that mixture of satisfaction and disappointment you get when stereotypes turn out to be solidly grounded in fact.

The equipment ranged before these photographers might not be familiar to their predecessors – the pioneers of nature photography over a hundred years ago – but their technique will be. They have come to this place knowing that a specific bird will disport itself for their photographic pleasure. Those early hobbyists, alive to the possibilities of the new technology for nature photography, adopted a similar approach. They would go to where they knew the birds were – usually the nest – and take their photographs safe in the knowledge that their valuable resources wouldn't be wasted.

But one person adopted a different strategy. Emma Turner preferred to set up her hide near a feeding station and wait patiently to see what turned up. This 'wait and see' method might not have had the reliability of the more widespread technique, but it produced more memorable images than the generic 'bird at a nest' type common at the time.

Photography back then was, to put it mildly, a palaver of the first order, but Turner went about it with a zest that would put many modern exponents of the craft to shame. Cheerfully enduring the hardships of bad weather and discomfort, she thought nothing of lying concealed on the ground for hours, her camera poking up through a layer of rotting vegetation, the better to photograph whatever might come by: 'Once or twice I felt the slender bill gently prodding my cheek all over, and once it was thrust into my ear . . . The rubbish-heap method of photography was absolutely exhausting, but it had lively compensations.'

This description of an encounter with a snipe, from her book *Broadland Birds*, written in 1924, sums up her approach, and the results were groundbreaking. Her photographs of young bitterns, taken in 1911 at Hickling Broad in Norfolk, weren't just confirmation that the bird was breeding in Britain again after an absence of fifty years – they were a slap in the face for the prevailing orthodoxy that proof of sightings could be obtained only with a gun. These photographs, and others like them, paved the way for the gradual exchange of one type of shooting for another.

How far we have come since then.

The procedure Turner would have used to take a photograph was, by today's standards, unthinkably tiresome. For one thing, the weight and unwieldiness of her apparatus would have made any photographic excursion physically demanding. And the hit-and-miss nature of exposure times, as well as the lack of any compensating features for light quality and so on, must have led to huge frustrations.

The camera was a box to which a lens was attached. You focused the lens using a focusing screen, which you then removed so you could insert the photographic plate. You exposed it for just the right length of time (you hoped), removed the plate, reinserted the focusing screen and started again.

If Emma Turner were around today, I would show her my phone with some sheepishness.

'Yes, I just, you know, take it out of my pocket and press the screen, and I can do that as often as I like, and, well, just look at this picture of a heron.'

This is the nature of progress. The quality of photograph I can take with my phone was unthinkable ten years ago. Go back another fifteen and the expression 'I've just taken a photograph with my phone' would earn you a blank stare. The cameras of my childhood are now museum pieces, yet even those would have represented unimaginable luxury to Turner. And what she was doing would have seemed almost like sorcery to Anna Atkins in the 1840s.

Turner's bittern photographs are, by any measure, remarkable, capturing the bird's intrinsic weirdness, its alien quality, to perfection. I would be more than happy to take a photograph half as good – so, I suspect, would most proper photographers, including the ones arrayed before me in an expectant crescent.

They appear to be playing the 'wait and see' game favoured by Turner, but in truth they know for certain that their vigil will be rewarded.

I walk on a bit. The path takes me slightly away from the trees, but now I can see what the photographers are looking at. A large branch sticking up from the ground. Perched on top of it, an indistinct grey shape. I bring my binoculars to my eyes.

Hello, Colin.

Colin the cuckoo has been coming to Thursley Common for at least six summers now. All the way from Africa to the exact same spot. He's achieved some celebrity with local birders, and especially photographers, because of his apparent love of the limelight.

Redstarts and stonechats and other heathland birds will come to the perch when he's not there, and no doubt they're welcome, but really they're just the warm-up acts. Colin is the main attraction. He comes and goes, but the consensus seems to be that if you wait long enough, and are armed with the necessary inducements in the form of mealworms, he will disport himself quite happily for a few minutes while you take enough photos to sink a battleship-shaped hard drive.

As I watch, he flies up, does a little loop, then flies back. Colin the Performing Cuckoo. Thrill to his acrobatics! Marvel at his skills!

Even more interesting is the visible awakening of interest among the circle of photographers. I'm too far away to hear them, but I can imagine the flurry of shutter clicks at Colin's every move.

I can see why they're there. You'll never get a better chance to see a cuckoo up close, to get pin-sharp photographs of one in all sorts of poses. And they've waited patiently for their reward. The ability to sit and do nothing for hours, just for that one perfect shot, isn't one I have. I'm too impatient, too keen to see what else is around.

I toy briefly with the idea of joining them, but I quickly dismiss it. No doubt they would be friendly and welcoming, but I'm in the mood for solitude, and with my small lens and general confusion about f-numbers and stopping down and filters and anything more complicated than fiddling around with the settings until it feels right, that circle is not the place for me.

Besides, I have the memory of my own cuckoo moment to keep me going through the long winter months.

I turn round and head back to the car. Colin will manage fine without me.

Back home. Instagram. Twitter. Scroll, scroll, scroll. I share my best lizard photograph. I don't have to do this. I could keep it to myself, but there are people on there who I know will enjoy it. I know from the occasional interaction that a small but significant number of my followers* have mobility problems, are unable to move far, or even at all, beyond their own homes. For them, photographs shared by people from all over the world are among the great pleasures afforded by the rise of social media. I know, too, that this enforced immobility leads a lot of them to be supreme observers of things,

* Really I'd like to call them 'acolytes', but I suspect that would be considered unseemly.

looking closely at the world in a way I, with my instinct for relentless busyness, find counter-intuitive. It's a decent photograph, too, capturing the lizardness of the lizard at close quarters. But there's also an element of ego, as there is with all social-media posts. By posting it I'm saying, 'Look at this beautiful thing', but also, 'Look at me, how clever I am to photograph the beautiful thing.' If I struggle with this aspect of social media, I'm also not strong enough to break free of it, so I scroll, scroll, scroll.

An image breaks the flow. It is, by any standards, strikingly beautiful. The composition is unusual, the subject viewed from a low angle, catching the setting sun behind it. And it's an action shot, capturing the speed and energy of the subject while somehow retaining astonishing clarity and focus.

The subject is Colin, coming in to land on his perch. The bars on his chest, the set of his wings, the fervent gleam in his eye – they're all in perfect balance, and given a warm glow by the setting sun just out of shot. It's an image to retain in your mind's eye. It brings Colin to life, captures his essential character, stands out from the dozens of photos I've seen of this much photographed bird.

I give it a 'like' and a retweet, because that's what you do. And I file the image in my memory banks, in a little virtual folder called 'cuckoos – pleasing images', where it rubs shoulders with my own mental images of a cuckoo fending off a whitethroat.

Perhaps all that hanging around was worth it after all.

5

WHEN NATURE CHANGES

In which our author yearns for times past,
tries to accept the inevitability of change, and
pays homage to the great John Clare

I remember the skylark most of all. I'd hear it before I saw it, the bubbling skirl of its endless song taking over from the engine noise as the 280 bus pulled away from the stop at the end of the lane. And then I'd look for it, scanning the blue sky for a distant quivering dot, rising and rising, apparently propelled upwards by the motor of its own sound.

Then I'd walk slowly home, hedgerows to left and right, cow parsley and nettles jostling for position in the ditch, and just before the bend in the road the trees arching over the lane, meeting in the middle, offering welcome shade and coolth in the heat of those endless childhood summer days of my memory.

It's faulty memory, of course – in my head the skylark sings alone, unsullied by the voices of the cars and trucks that will certainly have rumbled along the road with the bus, belching leaded fumes into the air; and the sun is always shining in this fantasy world, my memory conveniently deleting the days of squally rain that no doubt blighted my summer holidays and made the slow passing of time seem a burden rather than a luxury. But it's good to recall the fictional idyll, with its soundtrack of cooing wood pigeons, mellifluous blackbirds and the distant *ker-chook* of a pheasant as I whiled away happy hours pongling about the garden with a cricket bat and ball. The background to it all, in those high-summer days of somnolent heat, was the soft hum of insects, thousands of voices blending into one gentle burr, untainted by anything more disturbing than the occasional lawnmower and the contented burblings of *Test Match Special* on the radio. And then, as summer gave way to autumn, the excited chattering – the sweetness of the voices cut with a guttural quality – of swallows lining up on the phone lines in anticipation of their ridiculous, extraordinary journey to southern Africa.

Then one year I noticed a different note, barely but clearly discernible, clashing with the sounds of nature, jarring despite its quietness. I was a relatively unengaged child, but even I knew they'd been building it, and now it was here – a motorway, cutting a swathe through the countryside a couple of miles away, the convenience it brought undercut by the realisation that nothing would ever be the same again. This sound would always be there, a background drone I would have to get used to.

Progress has its drawbacks.

*

The crowds of starnels wiz and hurry bye
And darken like a cloud the evening sky
('Autumn Birds')

I could drive, but the train is cheap and quick, and Helpston not far from Peterborough. Besides, how better to get into the spirit than to explore John Clare's neighbourhood on foot? I have my route mapped out on my phone – as direct as possible to the outskirts of the town then following footpaths through farmland and copses and along a stream to the village – and the recent torrential rain has abated for just a day to deliver fine conditions for my walk.

There's a cockle-warming heartiness to autumn walks, especially when the leaves are doing their russet-and-gold thing on the trees rather than decaying into slimy sludge underfoot. The colours, the slight but not excessive chill, and the prospect of a log fire, a baked potato and a well-earned glass of wine, all combine to put a spring in the step.

The start of the walk, out of the station and through the back streets of Peterborough, has the plainness of suburbia everywhere. I shut out the traffic noise and concrete by focusing on every bit of greenery I can find. Ooh look a bramble. How lovely. Lovelier, in any case, than the railway siding behind it and the blank advertising hoardings alongside.

A weak late-afternoon sun slants through a mostly dull sky, a flurry of starnels* skedaddle over my head to perch on a TV aerial

* Starlings. So imbued am I with the Clare spirit that I start calling the birds by the names he would have used. Don't worry – it'll pass soon enough.

over the road, and the general drabness is enlivened by the bright red of rowan berries.

My pace is brisk enough, but it's not long, as I reach the outskirts of town, before any jauntiness in my mood is replaced by a sullen muttering. It's a situation familiar to anyone who has spent any time walking anywhere. All is well until it isn't. The network of footpaths threading through town and country serve the pedestrian well, by and large, but then you reach a moment when you're walking by a road and suddenly the footpath peters out and turns into an eighteen-inch strip of grass verge and before you know it you're in danger of being toppled into the brambles by the whoosh of displaced air as an Eddie Stobart lorry zooms past at 50 mph two feet away. My feeling that this is not a world built for pedestrians is confirmed as I realise that the only way to get where I want to go is to take my life in my hands, trot across the slip road as fast as I can and hope that I'm not mown down by a speeding Audi.

Maybe I have only myself to blame. Maybe I wasn't paying attention. There was probably a way round it that I missed, a small green sign saying PUBLIC FOOTPATH, or UNDERPASS, or COME WITH ME IF YOU WANT TO STAY ALIVE. But to go back and look for it feels as if it would be not just a pain in the arse but a waste of time too. I'm already aware that I need to walk briskly if I'm to make it to Helpston by nightfall. So I trot and I pause and then I trot again, and after a left turn and a right turn and another spell in the long grass I'm suddenly in open country and being mewed at by three puddocks.

*

I've seen bumbarrels on the wing
Full twenty flitting in a lot
And now and then on branches hing
Then peck and seek another spot.
('Birds Nesting')

He was productive, John Clare, susceptible from an early age to the disease of writing. He wrote over three and a half thousand poems, not even a quarter of which were published in his lifetime. And if among their number are poems of an epic nature like the Byronesque *Child Harold* and *Don Juan*, or satire in the form of the long poem *The Parish*, it's his affinity with nature for which he is best known.

The label attached to him by his publishers – 'the peasant poet' – does him no favours, carrying as it does the implication of simplicity and roughness, as if he were some sort of rustic genius risen unschooled from the land. But it helped establish his reputation, and if it was at best a distortion of reality, it set him apart from his more well-heeled contemporaries.

His circumstances were certainly different from many other poets. I don't suppose Byron was ever forced, by a shortage of materials, to attempt making his own paper and ink – the former by scraping thin layers of bark from a birch tree, the latter a homemade concoction of bruised nut galls, dyes and rainwater. And while he was feted by the London literary scene for a while, and made several visits to the capital at the beginning of his career, his home was Helpston, and it was there and in the surrounding countryside that his poetry lived. There is nothing grandiose about it, just straightforward observation and description of what he saw; and he saw a lot – small details of the everyday that others might not notice or think worth mentioning.

He would tell you about an easily overlooked bird, the carrion crow:

> I love the sooty crow nor would provoke
> Its march day exercise of croaking joy
> I love to see it sailing to and fro
> While feelds, and woods and waters spread below
> You're with him as he finds a nest
> on the almost bare foot-trodden ground
> With scarce a clump of grass to keep it warm

and share his innocent excitement at the emergence of the bird and its subsequent identification:

> – Stop, here's the bird – that woodman at the gap
> Hath frit it from the hedge – 'tis olive green –
> Well, I declare, it is the pettichap!*

If you're a birder, that process – stumbling across a bird and then identifying it – and the emotion it elicits will be familiar; if not, it's the kind of unalloyed pleasure that can overcome your initial scepticism and draw you in, and before you know it you own a pocket

* This could be any one of a number of warblers – it seems to have been a catch-all term. I've encountered it used variously to mean a chiffchaff, willow warbler, garden warbler or lesser whitethroat. It's also the name of a shop selling 'shirts for squirts' in Diagon Alley in the Harry Potter books.

field guide and a pair of binoculars and are looking for pettichaps in every bush.

If you've never heard a nightingale sing, then reading Clare's attempt to 'syllabise the sounds' out loud will give you an idea of it, or at the very least make you smile:

> 'Chew-chew chew-chew' and higher still,
> 'Cheer-cheer cheer-cheer' more loud and shrill,
> 'Cheer-up cheer-up cheer-up – and dropped
> One moment just to drink the sound
> Her music made, and then a round
> Of stranger witching notes was heard
> As if it was a stranger bird:
> 'Wew-wew wew-wew chur-chur chur-chur
> Woo-it woo-it' – could this be her?
> 'Tee-rew tee-rew tee-rew tee-rew
> Chew-rit chew-rit' – and ever new –
> 'Will-will will-will grig-grig grig-grig.'

And if you wanted to know what a bumbarrel's* nest looked like, he would describe it with all the accuracy of a field guide, and a hundred times the music.

> Of mosses grey with cobwebs closely tied
> And warm and rich as feather-bed within,

* Long-tailed tit. Black and white and pink and cute as all get out. Its many folk names included jack-in-a-bottle, hedge mumruffin, bottle tit, oven bird, fuffit and prinpriddle. But bumbarrel is the best of them.

With little hole on its contrary side
That pathway peepers may no knowledge win
Of what her little oval nest contains –
Ten eggs and often twelve, with dusts of red
Soft frittered

And if when reading his poetry I get the sense that he's taking me for a walk through the Northamptonshire countryside, showing me the commonplace, the everyday, stopping and waiting and looking and finding and looking again, I in turn would very much like to show him the puddocks that greet me as I leave Peterborough's rumbling metropolis behind and set out into the countryside.

Puddock circling round its lazy flight
Round the wild sweeing wood in motion slow
Before it perches on the oaks below
(*The Shepherd's Calendar*: 'October')

I should give them their official name, red kite, or even go for the scientific *Milvus milvus*, a pleasing tautonym. But 'puddock' is what John Clare called them, and the word has a nice pingy feel to it, so puddocks they are from now on.

They're quartering the copse and field to my right, their rangy, fingered wings flapping without urgency. But their languid movement is soon interrupted by the unwelcome attentions of two carrion crows, flappy black silhouettes making darting attacks, punching above their weight, getting close enough to bother them, to chase them away, but always somehow dodging a physical contest they know they'd lose. Their rough corvid *grawk* contrasts with the plaintive mew of the larger birds, which resist and dodge, forked tails flexing.

The territorial tussle holds my interest, and by the time the crows have seen the puddocks off over the woods and away to another territory, I'm fifteen minutes down on my schedule, and need to pick up the pace. The last pale gleam of sun pokes through the cloud of a late autumn afternoon, lights up the treetops in the copse at the far end of the field, flaming the russet and lemon tinges against the glowering backdrop of dark and heavy rainclouds. Muted, compared to the lurid display sometimes on offer at this time of year, but nonetheless possessed of a subdued charm. A soft mist rises from the ground, fields in bare autumn clothes of earth and stubble. If the surrounding countryside is flatter and the views less dramatic than I would normally be drawn to, it's doing a good job of making up for it with autumnal atmosphere. Jackdaws *chack*, magpies flare, a wood pigeon barges the air with its stout chest, fast white-barred wings aflurry. Two stout oaks, possibly just old enough to have been saplings in Clare's time, stand sentinel over the entrance to a field. From a hedgerow I hear the warning chirps and clips and stutterings of wrens and robins – 'Human! Looks harmless, but you can't be too careful!' A blackbird heeds the warning and bolts into low cover ahead of me with a staccato cluck.

Distractions everywhere.

A jogger comes towards me. He looks on the point of collapse, his hair plastered over his forehead, sweat dripping from every available part of him. But he still manages a breathless 'hello' to answer my uncertain greeting.

Subliminally spurred on by his athleticism, I pick up the pace, forcing myself to ignore the alluring chirps of a clutch of house sparrows* in the hedgerow to my left, paying scant attention to the piping contact calls of a family of bumbarrels in the branches above my head – not Clare's twenty, but just the six, behaving exactly as he wrote them a couple of centuries ago – and callously snubbing the winsome autumn song of a robin.

This landscape isn't beautiful or dramatic; it doesn't inspire lyrical thoughts. Clare himself described Helpston as 'a gloomy village', and while the surrounding countryside would in his time have been jumping with wildlife, now it is sadly denuded, an unending vista of flat farmland with few landmarks or distinguishing features. As such it's typical of a swathe of land that tourists mostly drive through on their way north, lured by the romance and dramatic vistas of the Peak and Lake Districts, or the rolling hills of the Yorkshire Dales.

But it was his, this landscape – or at least something distantly related to it.

> Enclosure came and trampled on the grave
> Of labour's rights and left the poor a slave.
>
> ('The Moors')

It's the old cliché: 'All this was fields when I was a lad.' Where I'm walking is still all fields, but in John Clare's youth it was more than that. The 'open field' system had the village – in this case Helpston

* Clare kept a pet sparrow called Tom when he was a child.

– at its hub, with strips of land radiating out from it. These strips were subdivided into 'fields',* 'furlongs' and 'lands', but the boundaries between them weren't marked with fences or hedges – the countryside was open, and rights of use were shared between landowners and commoners. Even the poorest had a sense of ownership.

Then came Enclosure. For Helpston and its neighbouring parishes, it started with an Act of Parliament in 1809, when Clare was sixteen. Its full force took a few years to bite, but by 1820, when he was twenty-seven, the landscape he had grown up with, had come to love and depend on, was transformed. The parish was divided into rectangular fields, and local landowners given free rein to further subdivide their land as they wished, marking the boundaries with fences and hedges, and excluding the common people. Furthermore, the former 'commons and waste grounds', which had provided grazing land and a valuable source of fuel for villagers in the winter, were similarly treated. Where before people could roam unfettered, they were barred by NO TRESPASSING signs.

It was all done in the name of efficiency, to make the land more profitable – but the human cost was ignored. By portioning off the land, and giving control of access to the landowner, the act drove a stake through the heart of the community. Whatever the economic arguments – some analyses show that the poor were in fact slightly better off after Enclosure – nothing can gainsay Clare's personal experience. These were formative years – he was making the transition to manhood – and for someone whose relationship with nature couldn't have been more intimate, whose poetry was soaked through with a sense of place, encapsulating the character of the local and

* Open arable land, rather than the meaning we now understand.

charting and recording its minutiae in exquisite and personal detail, the result was catastrophic. In effect it was a straitjacket for his soul.

Where Enclosure features in Clare's poetry, it imbues it with a bitter, sullen rage, a despair and regret for what has been lost. It changed his life, and the effects were devastating and long-lasting.

> . . . I had taken such a heedless obseverance of the way that lead over a cow-pasture with its thousand paths and dallied so long over pleasant shapings of the future . . . that twilight with its doubtful guidance overtook my musings and led me down a wrong track in crossing the common . . .

There comes that moment when you've been walking a while and you still have a while to go and you check the map and suddenly the route has sprouted an extra mile and a half you didn't realise was there and your walking shoes are sodden and dirty and probably a bit smelly and never mind the glories of nature, that hearty feeling is wearing distinctly thin and can you have that drink right now thanks very much.

No, you can't.

My walking shoes have served me well, combining the requisite sturdiness with extreme comfort. But in recent weeks a crack in the fabric has appeared in the left foot, and while their waterproofing was never absolute, all it takes now is a short spell walking through a wet patch for me to feel the ominous seeping warmth of growing dampness that presages a squelchy walk ahead. I've come to embrace this feeling, safe in the knowledge that any discomfort will be temporary. But just now I could do without it.

Other things I could do without include the two-hundred-yard stretch of mud in front of me, rapidly falling dusk and a phone battery at 1 per cent.

I'd been keeping track of my progress on the OS map on my phone, the convenience of GPS outweighing the more traditional charms of a paper map. The old-fashioned ritual of spreading an Ordnance Survey map on the kitchen table and working out your route in advance has been replaced by technological progress, and very handy it is too, except for the obvious limitation – when the battery goes, you're stuffed.

I use the remaining bit of battery to memorise as much of the route as I can. It looks pretty simple, a matter of lefts and rights and straight-ons, the grid system of fields enforced by Enclosure all those years ago working in my favour. I just need to remember where to go left, and where right.

The paths are mostly well signed, and I have a general idea of where Helpston is – over there somewhere – but a part of me wants to do away with the footpaths and fences and gates and stiles and not worry about whether I'm allowed to be in this bit or if I've gone wrong and am going to be confronted with a barbed wire fence at the far end and will have to retrace my steps.

Freedom is an illusion – someone, somewhere, is making rules.

The air is getting colder, thick clouds have rolled in, the light has almost gone, there is no moon, and it really is rather dark. My eyes will, no doubt, adjust to it, but oh dear here's a puddle. A muddy one at that. Distracted by the now extreme wetness of my feet and the rhythmic schlurping sound I'm making as I walk, I momentarily lose track of where I am. It really is a barren, featureless landscape, but through the murk I can see the looming shadows of a wood ahead of me. It suddenly occurs to me that this is Rice Wood, known to Clare as Royce Wood, one of his favourite haunts – 'the wood is sweet, I love it well' – and just to the south west of Helpston. I'm nearly home and dryish.

It is not Rice Wood.

This realisation dawns on me some twenty minutes later as I emerge from whatever-it-is wood in some confusion and almost total darkness. There was a moment, while stumbling around in it, the righteous path long since abandoned, that I thought I knew where north was again, but it was fleeting, and now all I have to guide me is the sound of a car rumbling past on a road some way off to my left.

A road. That'll do it. Find the road and follow it to Helpston. Easy.

A distant tawny owl hoots its approval of the plan.

I walk carefully towards where the sound came from. It's uneven ground, and I stumble a couple of times, but I'm making progress.

Another car goes past. I up my pace. The bumpy ground coalesces into a discernible path – smooth, tarmacked, welcoming. A silhouette emerges from the gloom – the silhouette of a tall gate, the kind of gate you can't climb over. Each side, an equally tall chainlink fence, guarded by a thick sward of trees.

I reach the gate and give it an exploratory tug.

It is, inevitably, locked.

> I am – yet what I am, none cares or knows;
> My friends forsake me like a memory lost.
>
> ('Lines: I Am')

There exists just one photograph of John Clare, taken in 1862, two years before his death. I see his large forehead, the straggly side whiskers, the quiet half-smile; but most of all I see an air of quiet benevolence, a gentleness, a man with whom you would want to spend time – the character shining through, even from behind the bushy eyebrows that half shield his eyes, even through the depths of confusion.

The depression from which he'd suffered for most of his adult life resulted in two stints in asylums. The first, in Epping Forest, ended after four years when he walked out one day and returned home (eighty miles or so) on foot, arriving, understandably exhausted and confused, four days later. His readmission to an asylum in Northampton a few months afterwards proved permanent. The asylum, far from the image it might conjure up to a modern person (especially one whose introduction to such places was through the film *One Flew Over the Cuckoo's Nest*), was a place of refuge, of safety. In both places he was afforded freedoms, able to walk freely into town and the countryside, and he continued to write until shortly before his death. But even though he was treated with care and respect, he felt incarcerated and unhappy, writing about 'the Hell of a madhouse' and 'captivity among Babylonians'.

The links between nature and mental health are nowadays well established and scientifically proven. Clare would have been able to point them out in a flash. It's almost as if nature – and in particular the countryside around Helpston – was inextricably connected with his mental condition. Even the move to Northborough, just three miles away, in 1832, had him yearning for the different landscape of his home village. And while he had suffered prolonged bouts of depression before then, they only intensified and lengthened as he grew older.

If there's little about the physical world today that he would recognise, there's every chance he'd find common ground with many people in the area of mental well-being. And while correlation doesn't necessarily mean causation, the increased acknowledgement of mental health problems has coincided with a time when disconnection from nature is also a growing concern.

There are other reasons. Of course there are. Mental health is enormously complex, and I'm not pretending that a walk in the park is going to miraculously cure everything. But for some people, that contact with the natural world gives them a daily lifeline, a thread to keep them connected in a disconnected world.

Not many of us will experience the intensity of Clare's relationship with his environment – to the extent that when it was taken away from him it felt as if he was being deprived of a part of himself – but anyone who has ever felt a moment of uplift at the sight of a tree or a bird or a mammal or any other thing of what my friend Stephen calls 'The Nature', or who has been shaken out of the doldrums by a walk in the park, or whose heart sinks at the prospect of concrete and metal and the constant smell and din of passing traffic, will at least have some inkling of it.

> Wrens cock their tails and chitter loud and play
> And robins hollow tut and flye away.
>
> ('Birds in Alarm')

It's relatively simple in the end. Retrace my steps as far as I can remember, then take the obvious path out of the woods, the one I missed earlier even though it was right there in front of me.

I spend the night at the Bluebell pub, next to Clare's cottage and where he worked as a potboy as a young lad. It has good wine, an excellent venison burger, and its warmth and cosiness are altogether welcome; and if I bring a whiff of the bog into the pub via my mud-drenched boots, they're polite enough not to comment on it.

The next morning I visit John Clare's cottage, using the audio guide selectively as I wander through the rooms where he lived with his parents, wife and six surviving children, and wondering at people's ability to live cheek by jowl.

As I emerge into the sunlight, a horse plods past on the other side of the road, well within the 20 mph speed limit. It brings to mind Thomas Bewick's 'improved cart horse', and is the kind of horse Clare would have known – large, grey, with a long floppy mane and an unhurried gait, somehow exuding an air of great patience and fortitude, and redolent of an age long gone – an impression slightly undermined by the ultra-modern hi-vis garb of its rider, a policewoman who answers my 'Morning!' with a friendly nod. What Clare would make of the vibrant yellow jacket is anyone's guess. The Helpston of today is littered with signs of the modern world – things that have stealthily accrued over the years and that we now take for granted: cars, trains, wheelie bins, electricity pylons, phone booths, level crossings, bus shelters, satellite dishes, traffic cones, meal deals at the pub, laminated sheets in the church porch with details of next week's harvest supper, electric gates, frozen meals, TWENTY IS PLENTY signs, churchyard mowing rotas, mobility scooters, discarded granola crunch-bar wrappers, burglar alarms, CCTV cameras, defibrillators, lottery adverts, sticky plastic banners on telephone junction boxes proclaiming FIBRE BROADBAND IS HERE, and convenience stores with posters advertising three for two on bottles of Vimto. Bewildering stuff, even to the initiated, once you start noticing.

But then it's impossible to predict what would seem strange about our present to some putative time-traveller. For all we know, he might take the cars in his stride but be bewildered by a manhole cover.

What he couldn't fail to notice, though, is the extent to which nature has been despoiled since his lifetime. If Enclosure was enough to disrupt his life, his reaction to the state of nature today hardly bears thinking about.

Perhaps his natural confusion and trauma* could be eased by showing him the things that are the same – the song thrush doing battle with a snail, the crows rowing their way across the sky, the bumbarrels and the puddocks and the crow-flowers too. And if that weren't enough (and I suspect it wouldn't be) then I would take him to the cottage of his birth and show him the quiet veneration and love for him and his work and I would say, 'There, see?', and it would be all I could do and I'd just have to hope that it helped.

I visit his grave. It's a small country churchyard, similar to the one twenty yards up the road from my childhood home. You can chart the story of a community from such places, reading between the lines of the simple epitaphs. If you look carefully, you can just make out the engraving on Clare's stone: SACRED TO THE MEMORY OF JOHN CLARE, THE NORTHAMPTONSHIRE PEASANT POET. Then his dates, BORN JULY 13 1793 DIED MAY 20 1864. And on the other side, A POET IS BORN NOT MADE.

* Not to mention the effects of time travel, which I'm told can really knock you for six.

The stone is blotched black, white and grey all over, the evidence of two centuries of weathering. Two patches of yellow lichen have colonised the 'R' of 'MEMORY', and everywhere there are little clumps of green moss sprouting from the crevices of the engraved letters, which have been rendered almost illegible by the passing of time and the accretions of nature. But the John Clare Society, with the love and pride common to all their activities, and aware of the needs of visitors like me, have erected a discreet and elegant pillar at the foot of the grave, etched with the same wording.

I sit on the bench a few yards away. A blue tit calls – *tsee tsee* – and from behind me comes an answering *chip* from a great tit. And then, as if sensing the mood, a robin starts to sing its autumn song, a more melancholy and downbeat affair than the spry, silvery spring version. We like to ascribe human emotions to these things – the goldfinch's song is cheerful, the curlew's cry plaintive, the barn owl's screech fearsome. Never mind that the science tells us that the robin's autumn song is territorial, and not, as we'd like to believe, performed for our benefit, to induce a melancholy tinge as the days shorten and the cycle of life reaches the decay phase – it's the anthropomorphism that counts.

Nevertheless, the sound does impart a mood, and I, already in the mood to be in a mood, am carried along by it. Difficult not to be struck by its timing; difficult not to succumb to feelings engendered by the occasion; difficult not to see it as a sign of something or other. Resolutely unsentimental, unmoved by faux spirituality, unimpressed by supernatural gubbins, I still allow myself to be quietly moved by the synchronicity of a bird choosing to sing at just that moment, a thread of connection spanning the centuries. You could write a poem about it, if you were so inclined.

That was how Clare expressed his love of the natural world around him – in the way most intuitive to him. And his words – like Thomas Bewick's images – endure for future generations, speaking for and to people who might not have either the inclination or means to express themselves in the same way. For many people, their relationship with nature is not only deeply personal, it is internalised. When we see someone looking at a tree, we have no way of knowing what's going on in their heads. Maybe they're silently composing poetry; perhaps they're wondering if they left the iron on; or they might just be thinking about the deliciousness of really good chips. It is, and should remain, a mystery. But sometimes the Thomas Bewicks and John Clares of the world see fit to record their reaction in the form of art, and that in turn affects people in different and unknowable ways.

I turn to watch the robin. It's perched on a cypress tree in a garden behind the churchyard, its silhouette crisp against the clear blue autumn sky. I admire it for a second. Then, master of bathos, it stops singing, does a little poo, and flies away, and the cloying sentimentality of the moment is mercifully tinged with earthy reality.

Four decades on, I revisit the village of my childhood, driving along the same motorway whose incipient rumblings clouded my summer afternoon all those years ago. If I notice the changes, they're at least partly offset by the things that have stayed the same. The skylark fields – also homes, back in the day, to flocks of lapwings billowing up behind tractors, fieldfares feeding in their droves during the winter, and grey partridge blending in with the stubble and loam – are a golf course. I navigate my way round unfamiliar roundabouts and slip roads, past a motorway service station and along what we used to call 'the main road' but now feels like a puny imitation of one. My impression is that everything is more crowded, fuller of the stuff of modern life. Part of this is the natural shrinkage of your childhood world brought on by adulthood; but perhaps it's also to do with my general impression of the countryside being stifled, given less room to breathe.

But the horse chestnut by the churchyard is still there, its spiky offspring dropping to the ground in autumn and scattering conkers to trickle down the slope towards our old house.

The extent to which my memory is faulty can be measured in the little misrememberings: this gate was further on, surely? And weren't there four houses there, not three? And there was definitely a copper beech by the war memorial. A handsome, well-proportioned tree of age and majesty, its maroon foliage giving way in winter to dark straggly limbs reaching for the sky.

There was. I'm not misremembering. But age and disease finally caught up with it, and it's been cut down only recently. It'll be replaced, and the cycle will begin again, and in a hundred years, with any luck, a twelve-year-old might just walk past it and up the lane, clutching a pair of binoculars and hoping to catch sight of a yellowhammer.

I park by the churchyard, and walk through the village, which, in accordance with the laws of ageing, seems 10 per cent smaller than when I was a child. Swallows and house martins by the dozen bounce and swoop around above my head. Flattering that they chose to mark my return with such a winning display.

I walk beyond the war memorial, past the beech-no-more, up the lane and left and right and left again and over the gate and through the fields, and if it's not the same as it was forty years ago it's doing a decent impression of it. And when I reach the river and see a heron I stop a while and really look at it, take in the Dumbledore beard and dagger bill, the avid eye and stilt legs, the grey cloak of long, straggly feathers down its back, the impression it gives of great age and wisdom – a false impression, given that all it's doing is waiting for a fish to move and provide it with lunch. You could invent this bird, make it up out of your head and draw it on paper, and people would look at it and nod and say, 'Yes. Good.'

And then, with a hoarse croak which only serves to reinforce its prehistoric credentials and remind me that in the grand scheme of things forty years is the barest speck of time, it rises and flies away on slow, heavy wings, and I take it as a hint that the meeting is over, and head back to the car. And when, as I stroll back across a wheat field, the summer sun warm on my skin, there's a flurry up ahead, and three birds, small and brown and scruffy, fly up and bounce across to my right with a *skrriddlup* and a *skrrrr*, it takes me a second to realise they're young skylarks, and for the most fleeting of moments I'm eleven again and it's as if nothing has changed at all.

INTERLUDE

In which our author indulges himself in a short rant on the state of the planet in the early twenty-first century, including (but not restricted to) matters such as climate change, single-use plastics, mass extinctions, the depletion of nature, rampant pesticide use, plummeting insect populations, cruelty to animals, wanton littering, pollution, trophy hunting, the burning of the Amazon, the inaction of governments, the innate arrogance and selfishness of humankind, our apparently limitless capacity to treat the other inhabitants of the planet with disdain and contempt, and our equally limitless inability to see, appreciate or care about the consequences of our own short-sightedness with regard to the environment; with observations on the feelings of helplessness engendered by all the above (and much much more), as seen from the point of view of a concerned individual, and expressed in the form of a mildly allegorical tale about an English off-spinner and a broken lawnmower

Once upon a time there was (and is – he lives still) a man. We shall call him John Emburey, for that was (and is) his name.

John Emburey was a cricketer, bowling off spin for Middlesex and England. He was also an occasionally useful lower-order biffer, and (even by the standards of the 1980s) a fairly unathletic fielder whose stock-in-trade, in the immortal words of John Arlott, was to 'fall in behind the ball and accompany it to the boundary'.

None of this is relevant – I merely add it as a bit of colour. Cricket-haters, agnostics and don't-give-a-toss-ers can safely ignore the last paragraph.

What we want to focus on are five words once spoken by John Emburey. History doesn't relate when he said them. That doesn't matter. It's possible he never said them at all, but that doesn't matter either. These apocryphal tales, embellished over the years, their origins lost in the mists of time, are often better than the real thing. The point is that in those five words are embedded Socratic levels of wisdom, the kind of thing you expect from philosophers, sages or statesmen – not, with all due respect to purveyors of that fine and much maligned craft, from English off-spinners of the early 1980s with a bowling average of 38.4 and a strike rate of a wicket every 104.7 balls.

Coincidentally, the five words are spookily relevant to the state we find ourselves in vis-à-vis the current viability of the planet in general, and the environment and natural world in particular.

As I say, the time and place of their utterance needn't concern us. Let's say it was Lord's cricket ground at some point in the 1980s. Similarly, of the circumstances of their uttering we need know only that they involve a broken lawnmower – a broken lawnmower stubbornly defying the ministrations of several members of the Lord's ground staff.

John Emburey, passing by, was asked for his input.

He examined the mower. He fiddled. He tinkered. He did all the usual things people do when trying to mend something. Finally, he stepped back with a laconic expression on his face. He chose his words carefully.

'The fucking fucker's fucking fucked.'

He might have been right, but (just to stretch the metaphor beyond breaking point) I'd like to think we are all capable of wielding a spanner.

6

NATURE ON DISPLAY

In which our author looks at nature through the medium of a variety of glass screens, and worships at the altars of Rothschild and Attenborough

The iguana stands still, alert, its head flicking briefly to the left, eye unblinking.

Something's up. You can tell by the music.

Racer snakes. Slim and long and writhy. Quite a few of them, sliding across the sand in search of prey. The iguana must stand completely motionless and hope the snakes don't see it.

They see it.

With a strange prancing gait, and to a soundtrack that wouldn't be out of place in *Game of Thrones*, it's off, the snakes in close pursuit. You can keep *The French Connection*; *Speed* was lightweight in comparison. If you want the thrill of the chase, look to the modern nature documentary.

Pounding drums and scurrying strings accompany the iguana on its fearful journey away from danger. The snakes – sinister killing machines – are the bad guys; the iguana – plucky, resourceful, fighting overwhelming odds – is the hero. The snakes get closer, gaining on the iguana, inexorably closing the gap. Even as a hatchling, recently emerged from its sandy birthplace, it can outstrip the snakes on the flat. But it's run straight into an ambush. More snakes await, and now they have it in their grasp, enveloping it in their slithering embrace. There are five, six of them – it's difficult to tell with all the writhing. The iguana struggles briefly, then seems to accept the inevitable. The jig, surely, is up.

But no. With a monumental effort, it wrenches itself free and bounds away from the despairing lunges of the snakes, up the rocks. It leaps across a gap, beyond their reach, to its parents and safety. The snakes must find another meal.

In the alternative narrative, the snakes are the main protagonists – driven by hunger and trying to take advantage of a rare feeding opportunity, ingeniously pooling their resources and working as a team to find and secure the food necessary for their survival. But we prefer iguanas to snakes, so the film-makers choose this version.

On a sofa in south London we watch, transfixed. The tension is almost unbearable. As the snakes exert their stranglehold I let out a tortured 'hnneeggh', and then, finding my voice, 'RUN, IGGY, RUN FAR RUN FAST THEY'RE GONNA KILL YOU!' Finally, as the scene comes to a close, the music relaxes its grip and the narrative shifts to the charming habits of the snares penguin (accompanied by appropriately winsome and playful music, naturally), we can breathe a little more easily.

Nature programmes weren't always like this.

Gone, apparently, are the days of David Attenborough standing next to an armadillo, genially explaining its sex life. These flagship BBC programmes – *Planet Earth, The Blue Planet* and so on – are high on budget, high on production values, and also, dare I suggest, very slightly high on their own brilliance. I applaud them for the sheer dedication and skill required to capture these images, for bringing the astonishing richness of the natural world so vividly into my Sunday evening; I also worry that, while entertaining and educating us, they unconsciously raise our expectations. Nature, in our daily experience of it, will never be this cinematic. And while it's important and amazing to be shown the glories and grandeur of natural phenomena of all kinds from all over the world, there's a danger that the definition of 'nature-lover' becomes 'one who loves nature programmes'. No matter how immersive and gripping these programmes are, they're no substitute for getting out into the pissing rain in the vain expectation that the drab mudflat before you will yield something more dramatic than a distant and bedraggled shelduck.

I realise I might not be selling the real-life nature experience in the most effective way. Let me have another go. I'll try to do it without sounding too earnestly evangelical.

If we are part of nature, and it is part of us (we are, and it is), then it follows that any disconnection from it means a disconnection from ourselves. This doesn't mean we all have to stalk the local common with a pair of binoculars, marvelling at every blade of grass. And it absolutely isn't to dismiss the importance and value of good nature programmes – as should be clear from the passage above, I love them as much as the next sofa turnip. But we should also beware of mistaking them for the real thing. They show us what are commonly thought of as the best bits – the drama, the action, the pitting of often spectacular and exotic creatures against each other or the elements. And having seen a pod of orcas take out a seal in the frothing spume in crystal-clear high-definition super slo-mo, when we venture out into the real world, the bedraggled shelduck might come as something of a disappointment.

But sometimes the best bits aren't dramatic at all. They can be found in the simplest and nothingest of things: the aforementioned blade of grass, if you like; or the chaotic tangle of a bramble in autumn, heavy with fruit ripe for the plucking; or that particular sensation you get when the wind picks up a fraction on a not-quite-warm-enough spring day, and brings a chill to your cheek in a way that reminds you what it is to be alive; or simply the undervalued pleasure of sitting on a bench and doing nothing, looking at nothing, just allowing everything, including yourself, to get on with the underrated business of existing.

I said I wasn't going to get earnest or evangelical. Sorry about that.

Anyway, if we think we need to travel to distant shores to see the visceral drama of snakes vs lizards, we need to think again. Snakes vs lizards plays out all around us on a daily and nightly basis. Spider vs fly, blackbird vs earthworm, sparrowhawk vs chaffinch, bat vs moth. A rich and varied franchise with limitless reboots.

Just don't expect it to be accompanied by a Hans Zimmer soundtrack.

Go back in time. Not far – just a hundred and twenty years or so, not even a blink in geological terms. To a time, at any rate, before television, before air travel, before every aspect of the natural world was freely available. To experience the exotic and outlandish – snakes vs lizards, or its equivalent – you would have to travel far. Africa perhaps, or Asia, or South America – journeys well beyond the scope of the ordinary person.

Or you could go to Tring.

Baron Walter Rothschild's vision was simple. He loved nature. He also knew a great deal about it. And he wanted to share that love and knowledge with everyone. Not just fellow zoologists and collectors and those in the know, but everyone.

He was uniquely placed to do so. Born into circumstances of extreme wealth,* he developed a love for nature at an early age. This manifested itself not just in the usual examination of beetles and worms and sticklebacks that might have been expected of any outdoorsy boy – indeed, when it came to collecting, Rothschild took over where others left off – but also in an unquenchable thirst for knowledge, acquired from books, his own observations, and his friendship with an amateur taxidermist who lived nearby. At the age of seven, he announced to his family that he was going to open a museum. Three years later, he did exactly that – a display of insects and birds and butterflies and all sorts of stuffed animals in the shed at the bottom of the garden. By the time Tring Museum opened to the public in 1892 he had assembled a collection of living and dead specimens that would have been the envy of any museum or zoo. And by 1937, when the museum and its contents were given to the nation, the collection comprised over 2 million butterflies and

* It was his father, Natty, who was responsible for the 'no less than two acres of rough woodland' quote.

moths, 300,000 bird skins, 200,000 bird's eggs, and similarly bewildering numbers of mammals, birds, reptiles, fish and invertebrates, stuffed and mounted for public display.

Rothschild had a keen eye, an encyclopedic knowledge of the natural world, and a big budget. Most of all, he was a relentless collector of specimens, both living and dead. It was this juxtaposition that formed the heart of Tring Museum. Inside, a collection of stuffed animals the like of which had never been seen; outside, the living collection, either roaming free in the 107-hectare park or kept in enclosures closer to the house. Wallabies mingled with rheas; cassowaries rubbed shoulders with emus; wild turkeys, spiny anteaters, dingoes, cranes, zebras, a capybara, pangolins, deer, giant tortoises and even a tame wolf – all brought together by Rothschild's breadth of vision.

But this wasn't just the whimsical collection of an eccentric millionaire. His reputation as a zoologist was already established, and grew throughout his life – Tring was a fundamental resource for scientists as well as a thriving visitor attraction for everyone else.

At its peak Tring received 30,000 visitors a year – no mean feat for an attraction on the edge of a small town badly served by public transport, at a time when private car ownership was not yet widespread. The key to this success was Rothschild's understanding of what people wanted. The standard museum experience was, in comparison, sparse, dry and exhausting – a forlorn traipse round endless corridors. It wore people down. Tring offered them stimulation – drama and excitement and all kinds of crazy stuff. Part of it was about the extremes of the natural world that remain fascinating to the general public today – the biggest, smallest, deadliest, wildest – but it was also to do with how the specimens were displayed. Rothschild insisted on arranging the display cases himself, and was able to see them through the eyes of the fabled Man in the Street rather than the keen zoologist. Thirty stuffed birds arranged in a row might become monotonous; put them all in one case, cheek by jowl but still all visible to the gaping punter, and they become a fascinating spectacle. And if you could see all that and then go outside to feed a cassowary or pat a zebra, a trip to the Chilterns suddenly took on a new, exciting aspect.

The most familiar photograph of Walter Rothschild shows him, besuited, top-hatted, and holding a piece of lettuce on the end of a stick, astride a giant tortoise.* It is a striking image, not least because of the size of both protagonists. Rothschild was a large man – six foot three and twenty-two stone – but devastatingly shy, and there's a touching ungainliness about his posture on the tortoise, as there is in photographs of him in human company, eyes consistently downcast. The habit of riding giant tortoises like seaside donkeys wasn't the only mark of his eccentricity. When he went to Cambridge he took a flock of kiwis from Tring because he couldn't bear to leave them behind. Another photograph shows him sitting atop a carriage, reins in one hand and long whip in the other. It's a perfectly normal picture for the age, except for one striking detail: the carriage is being drawn not by horses but by zebras, specially trained by Rothschild for the job. Having trained them, he drove the carriage up The Mall and into the forecourt of Buckingham Palace.

Despite these manifestations of eccentricity, which could so easily become tiresome, I find myself strangely drawn to Walter.

* There would quite rightly be a hue and cry about animal cruelty if such an image were published today, but *autres temps, autres mœurs* and all that. We judge the past by the standards of today with caution.

A man of his position could easily have kept his collection to himself, but in sharing his knowledge with both the general public and fellow zoologists he showed a certain generosity of spirit, a spirit still evident when you visit the house, now the countryside arm of the Natural History Museum.

Today there are, disappointingly, no cassowaries in Tring. Not alive, anyway. There is a buzzard, wheeling over the car park, letting forth its plaintive mew into a clear blue sky. On the path to the museum a worm wriggles to escape from a blackbird's beak, with inevitable consequences. In the entrance hall a teacher tries to quell the excitement of twenty of her charges, with an equally unsuccessful outcome. The children are milling, as juveniles of this most intriguing species are wont to do, and they are doing it in such abundance that for a moment the narrow hallway seems to me a roiling ocean consisting entirely of children. They're given pencils, clipboards, sheets to fill in. An old man, caught in the throng, displays nimble footwork to avoid a small boy careering towards him, only to come a cropper against a slightly larger boy careering in the opposite direction. He accepts his fate with a resigned smile and a raised eyebrow, and continues patiently against the relentless flow like a salmon working its way upstream.

And then, as if with the flick of a switch, the children are gone, dispersed to one of the back rooms, there presumably to fill their sheets with answers to questions about hummingbirds and quaggas and how many tortoises they could find, and I'm alone with over three hundred birds, dozens of smaller mammals of various descriptions, and a polar bear.

The mixed emotions I always feel when in the midst of such displays are intensified by the density of the specimens. The glass cases are close on both sides – there's just room for two people to pass each other – and reach from floor to high above head height. The presentation of the specimens, it seems to me, is as close to the way it was in Rothschild's day as possible – certainly it would be a struggle to cram any more animals into the cabinets.

It could be a morbid experience, this immersion in death. These animals, an advertisement for the taxidermist's art, occupy a strange hinterland – definitely dead, but preserved in lifelike poses, to give the impression that they've been frozen mid-action. Somewhere else in the building, away from the public gaze, are many more specimens, preserved in less histrionic poses and laid in drawers, their role in death not to inform visitors but to help scientists. These animals, crowding in on me from behind glass, are more alive than a painting, but also somehow less so. My mind, overwhelmed by their abundance and proximity, makes a weird leap, envisaging the glass cases melting away, and the animals slowly coming to life. It shies away from the inevitable ensuing carnage, but not, I suspect, before it's stored up a disturbing moving image for later retrieval at 3.46 on some dark and sleepless night.

The lifelike poses of some of the animals only contribute to this dark fantasy. A bat-eared fox stands on a shelf just above head height, in a defensive pose, apparently caught mid-yelp; opposite, a golden eagle dwarfs the other birds of prey, wings outspread, poised to launch; below, a capercaillie looks up reproachfully as if I've accidentally stumbled into it, its tail feathers spread in a warning to potential predators;* in the cabinet opposite, a gaggle of

* Either that or it really fancies me – I don't know which thought is more disturbing.

lemurs cling to branch stubs, wide eyes unanimously open in a silent chorus of startled reproach. Horror films have been made from less.

For the enthusiast, the close juxtaposition of related species offers an opportunity to make comparisons, like a 3D field guide; for the casual visitor it's an introduction to the diversity of mammalian and avian species on the planet – the merest snapshot of the tiniest portion, in reality, despite the apparent wealth of specimens in the room. But for most visitors, now and in Rothschild's time, it represents far more than they would see in the wild in a lifetime.

It's unnatural, of course – you'd never see even half these species in the same place and in such rapid succession – but also stimulating in a quite different way from the heightened reality of television spectacles. The frozen nature of these animals, the permanence of the pose or expression chosen by the taxidermist, nonetheless allows scope for imagination.

As Rothschild understood, people are naturally drawn to the dramatic. The most commented-on exhibit in this room, as people begin to trickle in behind me, is the polar bear, seated on its hindquarters, mouth half open, its bulk and physicality apparent even when denuded of life. While its familiarity as an icon of the fragility of the natural world means I spend some time looking at it along with everyone else, I'm also drawn to the out of the way, the unconsidered specimens lurking at the outer edges of the displays. The statuesque magnificence of a shoebill catches the eye from the back of a cabinet devoted to large stabby birds, the bulk of its eponymous appendage striking in both its size and the shape it lends the bird's face. The allure of this bird is partly due to our attraction to animals that look as if they're smiling (see also dolphins), but also to the bird's extraordinary general appearance – a contemporary reminder that even today we walk among dinosaurs.

A cabinet devoted to albino birds has a perverse allure. These aberrations were a particular interest of Rothschild's, and as I look at a puffin – colourful beak and black wing in place as usual, but otherwise completely white – I sense a slight disturbance in the matrix, as if I've been allowed a glimpse into a parallel universe.

By the door, a cabinet of small birds, none of them found in Britain. My knowledge of world birds is poor, but the first to catch my eye happens to be a familiar one, if only because the family it represents – their appearance, their behaviour, their exoticism, their everything – exert a particular fascination. It's a male Wilson's bird-of-paradise (*Cicinnurus respublica*). In real life it's a bird of spectacular beauty, a flying advertisement for the evolutionary success of primary colours: a pale blue head,* bright yellow nape and red back, framed by a dark chestnut torso and wings – Fisher-Price's My First bird-of-paradise. Add to that an iridescent green breast shield that flares out when the bird is displaying, purple legs and twin-curlicued tail feathers, like an extravagant handlebar moustache, and this starling-sized bird does all that's required of it in the 'gaudy exotica' department.

Needless to say, you won't see it in Britain; you won't see it anywhere except two Indonesian islands, Waigeo and Batanta. Its presence in this display, along with dozens of other far-off imports, is the legacy of Rothschild's wide-ranging obsession with collecting – anything and everything, from anywhere and everywhere, gathered

* The blue is the colour of the bare skin, not feathers.

under one roof and presented to the public to instil a sense of the wonder he felt in the face of nature.*

The colours on the specimen in front of me have been subdued by a combination of the taxidermy process and time, but it still projects something of the character of the real thing, the eye-catching tail feathers jauntily offsetting the downward swoop selected for it by the taxidermist. I imagine it in life, its call resonating through the Papuan canopy in much the same way as the excited babbling of the children, fact sheets presumably filled in, now resonates through the gallery. Keen not to be swept along by the ceaseless torrent, I skip a room and head upstairs.

I nearly walk past it. It's not exactly hidden, but tucked away at the top of the stairs, the kind of thing it's easy to overlook, assuming that if it's in a corridor it's not interesting.

It is interesting.

Hummingbirds. Lots of them.

If my fascination for birds-of-paradise is strong, it's dwarfed by my interest in these tiny miracle birds. Everything about them is amazing. Their size, their delicacy, the beauty of their plumage, the extraordinary adaptation of their bills, their tongues,† their flight, I mean have you SEEN THEM? They can hover, go backwards, forwards, sideways and diagonally, their wings moving too fast for the human eye, executing a uniquely intricate wing-stroke enabled by

* He was by no means a conservation saint, though. He was an enthusiastic duck hunter all his life – an interesting contradiction common to naturalists of the nineteenth and early twentieth centuries.

† Only recently, with the advent of high-speed photography, have we worked out that their tongues flick in and out of flowers up to twelve times a second, enabling a steady flow of nectar.

flexible shoulder joints, which means the wing can change its angle by nearly 180 degrees and they can generate lift on both down- and up-strokes. In the aerial-competence department, they're up there with dragonflies. Is it a coincidence that I'm equally fascinated by the hummingbirds' cousins, the swifts, with their similar mastery of the air? No. No, it isn't. We are attracted by what we can't do, so any right-minded human must surely be drawn to the mysteries of flight.

The hummingbirds in the cabinet have been arranged lovingly, some in a semblance of flight, but mostly perched, poised for action. Their names reflect the magical qualities they invoke: white-bellied woodstar, violet sabrewing, rainbow starfrontlet, fork-tailed woodnymph, black-eared fairy, white-vented plumeleteer. The iridescence in their plumage has mostly been preserved by the taxidermy process, but, like the Wilson's bird-of-paradise downstairs, there is something missing. In the case of the hummingbirds it's something even more central to their allure: life. Much as I admire the dedication of the people who put together these displays, a motionless cabinet of these little jewels doesn't even begin to do them justice. They're creatures from another world – unless you live where they're a part of life and not an impossibly distant slice of exotica.

I wonder how it would be to live in Costa Rica – home to more than fifty species of hummingbird. I'd like to think familiarity wouldn't lessen their impact, that I'd spend my life in a state of slack-jawed wonder. And then I remember the kerfuffle a year earlier when a robin – the gardener's friend, Britain's de facto national bird, doyen of Christmas cards in perpetuity but actually a right bastard when you get to know it – turned up in Beijing. Beijing is not on the robin's 'frequently visited' list. It's some six thousand miles out of its way. Its appearance there prompted a flurry of activity in the

Chinese birding community, and gave newspapers an opportunity to dust off their favourite 'twitchers flocked' headline. The sight of this most familiar bird in most unfamiliar surroundings, with a phalanx of photographers in the background, was a vivid reminder not to take the commonplace for granted, to look at the normal more closely, to appreciate the magic of the everyday.

Everything is rare somewhere.

A full tour of the galleries, examining each specimen closely and reading all the associated literature, would be more than a day's work, but this isn't how the casual visitor consumes a collection. Most people set their pace to 'saunter', preferring the gentle drift to full immersion. The eye is caught by the dramatic or the unusual or the familiar – the second most commented-on exhibit in the downstairs gallery was the red kite, familiar to locals as the most visible and dramatic bird of prey in the area – but a lot of the specimens are subsumed into a general mishmash of museumery. I suspect I'm the only person that day (or possibly any other day) to spend ten minutes making a close comparison of two nondescript wading birds – the green sandpiper and wood sandpiper – this being the closest I'd ever got, or was ever likely to get, to either of them.

Upstairs, it's a similar story. The balcony overlooking the downstairs gallery is lined with glass cases, and again they're packed to the gunwales with specimens. If I barely glance at them, it's not because I'm averse to fish, merely because I've gorged myself on birds and mammals downstairs and am in a benign state of satiation.

But my curiosity is piqued by the line of cabinets opposite the display cases, their contents protected by wooden doors, positively inviting you to grasp the handle to reveal a beetle OH MY GOD LOOK AT THE SIZE OF IT.

It's big, no doubt about it. For a beetle, anyway. I compare it with the length of my index finger and come to the scientific conclusion that it's at least 'huge' and probably even 'ginormous'.* The label tells me its name: *Acrocinus longimanus.*† A South American species from the family *Cerambycidae.* It dominates its display case to the extent that the other specimens pale into insignificance. Not only is it large, it's dramatic too, with long antennae and even longer front legs – the delicate curve of the former offset by the angularity of the latter. I examine the subtle beige-black marbling on its body, and wonder just how sanguine my reaction would be if I discovered one of them behind my hotel curtains, and through my fog of absorption I become aware of a palaver to my right. A flurry of human movement and a quiet but ear-catching groan. I turn, eyebrows raised. Two women, one my age-ish, the other very much not.

'I'm sorry,' says the older woman, 'but I just can't.' She's almost back-pedalling into the display case behind her, shoulders hunched, flinching from the sight of the beetle. 'I can't even look at them. I know it's ridiculous.'

I close the door on the beetle and give her a smile, thinking it perhaps best not to mention that if she turns round she'll come face to face with a sea creature whose gormless grin I find much more sinister than any insect, however large.

* I look it up later. Its body is about three inches, with forelegs even longer.

† The second part of the scientific name gives a clue – 'long forelimb'. The English name, 'harlequin beetle', is less scary.

'I know I shouldn't be scared, but it's just the way I am. Anything creepy-crawly just gives me the . . .' I'm silently willing her to complete the rhythm of the sentence with the word 'heeby-jeebies', but she leaves it unfinished, its meaning clear. 'And I know I've passed it on to her.' She makes a strange sideways gesture with her head towards her daughter, who gives a little nod of assent.

'We're both trying to be better about it, but . . .'

The fear is left hanging in the air, unaddressed.

I find the psychology of it fascinating. Rational acknowledgement of irrational fears, grist to the hypnotist's mill, no doubt. But in particular it's the unconscious transmission of terrors and prejudices to one's children that strikes a chord. I have sympathy with both of them. While I react with curiosity and fascination to these insects, there are sound evolutionary reasons for our instinctive fear of things that might kill us, or even things that just look as if they're related to things that we somehow believe might kill us. Even if they're long dead and pinned to a bit of white cardboard.

Keen to respect their feelings, I wait till they've moved on before I open the next case, which is just as well, because it contains a spider that, at first glance, appears to be the size of a small pony. It would be dishonest to pretend that my first, ungovernable instinct isn't to recoil just a bit, but, rationalist to the core, I quell this instinct and take a moment to admire the spider's spideriness, before closing the case gently and making my way upstairs, there to spend quarter of an hour in the relatively safe company of a roomful of dead horses.

Fascinating as the displays – both animal and human – have been, I've had enough. Too many dead things. Too many glass cases. If only to balance it all out, I need the real thing. Luckily, Tring has that as well. Step outside the museum and you find yourself in an attractive park – woods, chalk grassland, wooded pastures and unfarmed wild meadows.

It's warm and bright and fresh, a light but steady breeze brushing away any cobwebs accumulated in the dry atmosphere of the museum. The buzzard has gone, the blackbird too. The path leads from the car park over the A41 via a bridge, and into the park. I take my time, sauntering through the pasture towards the wood, trying to imagine this place of magpies, blue tits and squirrels as it was in Rothschild's time, picturing the incongruities of loose wallabies, cassowaries, emus and rheas in the Hertfordshire countryside. I try, too, to put myself in the mind of someone for whom that kind of thing would be entirely normal. I fail.

In the absence of wallabies, the only large animals in evidence are a group of cows huddled under an oak in the middle distance, but they're singularly uninterested, looking at me with slow eyes and bovine diffidence as I walk past at a respectful distance.

A walk in a new place always carries with it a sense, however mild, of discovery. We might know roughly what kind of walk it's going to be – in this case, I'm helped by a map of the area which offers an array of options of different lengths – but the details remain unknown.

Will there be views? Will there be a pleasing selection of wild flowers specifically designed to highlight my inadequacy at identifying them?* Will there be tangled undergrowth, out of which pokes a small clump of bramble with incipient berries, all

* Cow parsley and rosebay willowherb – I can do those.

the better to remind you that the seasons are constantly changing and autumn starts sooner than you think? Will there be long-tailed tits flitting around out of sight above you, *tseep*ing to each other in their hyperactive way? Will there be dappled sunlight filtering through a high canopy to cast its rays on the floor in a way that makes you walk just that bit more slowly for fear of disturbing the spell? Will there be a good solid fallen tree trunk – just the kind you can sit on for a breather? Will there be a robin perched on a branch just a couple of metres above your head, singing a song so directly at you that you flatter yourself into thinking it's singing just for your benefit?

Will there be a kestrel?

On this occasion there will in fact be all of those things.

It's not one of the Great Walks, those memorable excursions etched into the memory because of their length or some momentous event or just the happy confluence of friendship, scenery and the intangible 'other' that might elevate it to the pantheon. It's just a good solid walk, awakening, as Beethoven so correctly put it in the subtitle to the first movement of his 'Pastoral' Symphony, 'cheerful feelings on arrival in the countryside'.* The kind of walk you could take home and introduce to your grandparents. And specifically on this occasion it acts as a pleasing counter to the quiet somnolence of the museum. The climb is stiff enough to get me out of breath but not exhausted; the woodland offers just the right mix of bosky excellence and avian temptation; and there is, without wishing to sound too curmudgeonly about it, a pleasing absence of humans.

* He did it in German, obviously: *Erwachen heiterer Empfindungen bei der Ankunft auf dem Lande.*

Oh, and there's an excellent view, the surrounding countryside stretching in all directions and naturally framed by the straggling branches of an oak. I'm mildly vexed by the absence of red kites, but it would be churlish to complain. Some walks are too short, leaving you thirsting for more but necessarily curtailed by the tedious realities of life; others are too long, the rain setting in at just the right moment to inflict misery;* this is a Goldilocks walk, and I return to the car park invigorated and satisfied by its just-rightness.

And if I secretly bemoan the absence of even a single red kite, I have to remind myself that nature isn't at our beck and call. If you want the animals on tap, you can get that on the telly.

An older programme. David Attenborough, standing next to a bird, talking to camera. The bird, perched on a branch, is astonishingly beautiful. The greater bird-of-paradise (*Paradisaea apoda*). All Attenborough wants to do is tell us about it: its two-tone cream and green head, warm brown body adorned with exquisite, gauzy, yellow plumes; its courtship display, wings held in an elegant arc as if shielding a child from the rain; its strident and resonant call, designed to draw a female's attention from afar; the history of its scientific name – 'bird-of-paradise without legs' – stemming from a time when Europeans had seen only wingless and legless skins of the birds imported for trade.

But the bird's having nothing of it.

'This, surely . . .'

'*WAUKWAUKWAUKWAUKWAUK – KWHOOR KWHOOR KWHOOR.*'

* I have learned to love a rainy walk, but there are limits.

'This, surely . . .'

'*KAAKAAKAAKAAKAAKAAKAAKAAKAAKAAKAAKAA*.'

'Of course, by the eighteenth century, naturalists realised that birds-of-paradise did have legs. Even so . . .'

'*KAAW KAAW KAAW KAAW KAAW KAAW KAAW KAAW KAAW KAAW KAAW KAAW*.'

Eventually, even the great man has to acknowledge that he's met his match. He nods, powerless in the face of the bird's determination to have the last word.

'Very well.'

Very well, indeed.

7

CAPTIVITY AND FREEDOM

In which our author goes to the zoo, dabbles in ethics, and, thanks to Sir Peter Scott, meets a Very Good Swan

There's a fracas over the tapir enclosure. Not that the tapir's taking any notice – he seems so disinclined to do anything he might as well be stuffed. I've admired his black and white colouring (evolved as a camouflage tactic), his body (somehow both delicate and lumbering at the same time), and in particular his long, fleshy, prehensile nose, with its mesmerising flexible top lip – so useful for those times you just need to get one more leaf into your mouth. He is a fine and admirable beast, but the fracas distracts me for a second.

The fracas is the responsibility of about twenty black-headed gulls. They are, when the fancy takes them, magnificently acrobatic birds, and their swoops and tumbles are accompanied by a frenzied squawking. Something's got them excited, but whether their squawks are a result of gullish exuberance or because they're actually having a barney, it's difficult to tell. Nevertheless, their display offers momentary distraction from the crazy antics of the tapir. Not that the crazy antics have started yet, but they're about to, any moment, I can tell.

The tapir is one of nearly six hundred species kept at London Zoo. It's an impressive list, ranging from corals to gorillas, and encompassing such snappily named beasts as the giant banded tailless whip scorpion (*Damon diadema*), the scarlet lady cleaner shrimp (*Lysmata grabhami*) and the giant thorny walkingstick (*Heteropteryx dilatata*). For all this diversity, though, most visitors are drawn to the glamour species – lions and tigers and bears* and so on – so I follow the crowds towards the gorilla enclosure (or Gorilla Kingdom, as they have it, which does make it sound more enticing for the gorilla).

I have no particular viewpoint on the gorillas.† They are magnificent, obviously, and while I'm tempted to snark about how much more magnificent they'd be in their natural habitat, that's the easiest of swipes to take at a situation which, as always, has more nuance and complexity than plain right and wrong.

When it comes to ethical arguments, my logic circuits habitually chase their own tails in an endless effort to reach a definitive conclusion. Animals shouldn't be kept locked up, but on the other

* There are no bears at London Zoo, which is probably good for the bears, but not so good for my analogy.

† I do, however, take great pleasure in the knowledge that their scientific name, *Gorilla gorilla gorilla*, is one of the very few tautonymic trinomials – scientific names where all three words are the same. Other examples are *Bufo bufo bufo* (European toad), *Giraffa giraffa giraffa* (South African giraffe), and, superbly, *Francolinus francolinus francolinus* (western black francolin).

hand these animals are endangered in the wild and the organisation that has them in this enclosure does extraordinary work around the world to help endangered species, but on the other hand surely they need more space to roam, but on the other hand the people charged with their care take great trouble to provide a suitable habitat, and besides, they know much more about that subject than I could ever hope to, but on the other hand don't they look miserable, all slouchy and glum, but on the other hand that's lazy anthropomorphism, perhaps it's just having a bit of a rest, and look at that one, chirpy as all get out, but on the other hand . . . and after a few rounds of this, I end up with about forty-six hands.

The easily disappointed* are quick to find fault with the zoo experience, citing the expense, empty enclosures, and the general failure of the animals to put on a show for them. But my fellow clientele on this bright spring day don't seem disappointed. There are twenty of them immediately around me, and their noise levels and excitability give the black-headed gulls a run for their money. For a moment I wonder what evolutionary adaptation makes me particularly suited to attracting large groups of small children while walking round zoology-based visitor attractions, but I quickly realise that such things are merely an occupational hazard. In any case, if the gorillas are, for the moment at least, not playing ball, there's plenty in the behaviour of the children to interest the budding anthropologist. There is, naturally, a fair amount of milling going on, along with more running than their teacher can handle. But four boys stand still next to me, transfixed by the gorilla nearby. It stares vaguely towards us for a few seconds, has a bit of a scratch, turns its head again, then

* You'll find plenty of that abundant human subset on TripAdvisor.

gets on with the fine art of sitting still. A gaggle of starlings hustle around on the floor near it. Like the black-headed gulls, they are interlopers from the wild, free to come and go as they please, unrestrained by cages or clipped wings. But they're canny blighters, these birds. They know where the free food is. The gorilla, for its part, is monumentally unbothered by them.

If the gorilla is notable for its stillness, the same can't be said for the boys. Trying to catch its attention, they jump up and down, waving their arms and making 'oo-oo' noises. One of them calls out. 'Hey! Gorilla!'

Nothing doing. The gorilla continues with its mooching, a master of the art, in the face of fierce provocation. I suspect it's had a lot of practice.

Three of the boys tire of their game soon enough, and dash off to torment a Bactrian camel instead. But the fourth stays for a few seconds, standing still and looking at the gorilla in silence. He seems to be looking at it for the first time. It's worth looking at, a living statue of dark, bulky nobility. The boy drinks it in.

Then, to nobody in particular, he says, 'It's sad', before dashing off to join his friends.

It's not clear whether he's referring to the specific emotional state of that individual gorilla or the prevailing situation worldwide. Either way, his willingness to do a bit more than goad the gorilla into putting on a show gives me a sliver of hope. Better for a child to see a gorilla and feel its sadness than not to see a gorilla at all. And if Gorilla Boy makes my mood a shade more buoyant, it's upgraded to 'perky' by Penguin Girl.

There's something about penguins humans find irresistible. Perhaps it's that they stand upright like us; perhaps it's the impression they give of noble endurance in extreme conditions; or perhaps it's the contrast between their comical incompetence on land and their mastery underwater.

Whatever the reasons, the penguin beach at London Zoo* is popular with children and adults alike. Its occupants, Humboldt penguins (*Spheniscus humboldti*), are declining in the wild, their official status Vulnerable. Here, these captive bred birds are safe from the attentions of guano-gatherers, oil spills, seals and sharks. Whether they benefit from the open-mouthed adoration of the paying public is debatable, but they get it anyway. They know how to put on a show, and the penguin beach has been designed to maximise the quality of viewing conditions. You really can get close and examine them. I spend a few minutes admiring their ease, speed and grace in the water. A couple of yards to my right, a small girl does the same. Her mother, behind me, apparently unengaged by the penguins, is chatting to a friend. The girl is unimpressed.

'Mum.'

Chatting continues unabated.

'MUM.'

Still nothing.

'Look! Mum! Look! LOOK!'

Third time's a charm.

'What is it, darling?'

'PENGUINS!'

The mother can barely mask her indifference.

'Yes. Lovely.'

* Which, because of reasons, we now have to call ZSL or something.

I'm firmly on the girl's side, and come within a toucher of turning on the mother. Because yes. Look. Penguins. But I hold my tongue, and the moment passes.

On the one hand it's dismaying. The mother has paid handsomely for this trip; it seems perverse to ignore one of the main attractions. But then I remind myself that the main thing is they're there. And even if the girl's fascination with the penguins is temporary, the image of their tubby black and white bodies is now in her head somewhere, added to the tumbling mass of images and sounds and thoughts and experiences that shape our understanding of the world. Given the ever increasing competition for our attention,* this counts as a victory.

A trip to the zoo remains a popular way to spend a morning. More than 1.25 million people visit London Zoo every year, and on today's evidence at least 1.249 million of them are under the age of thirteen. Whatever the reasons for their presence – whether it's a school trip, a way of keeping the kids occupied for a couple of hours while you catch up with a friend, or a genuine interest in conservation – engagement is better than complete unfamiliarity. When you meet an animal, stand close to it, observe it, maybe even interact with it in some way, your connection with it is greater, and it follows that you're more likely to want to protect it.†

* At this point you might think I'm going to have a pop at screens. I'm not. Screens are fine. I love screens. I've learned new things, met new people, had experiences I wouldn't have had without screens. But, like chocolate, alcohol and buying books you'll never have time to read, you can have too much of a good thing.

† The same goes for plants, by the way, as anyone who witnessed my fury when a stand of relatively scrubby trees down the road was cut down for a new development will understand.

Zoos have come a long way since the formation of the Zoological Society of London in 1826, and the spread through Europe and America of zoological gardens – intended primarily for scientific study – over the next fifty years. Once they were opened to the public, the remit expanded to entertainment, education and enlightenment (not necessarily in that order), and these priorities didn't give way to the less human-centric concerns of conservation and ecology until the 1970s. Despite the earnest efforts of zoos to big up their conservation credentials, with captive breeding programmes in zoos and extensive work protecting endangered species in the wild, you don't have to look far to find people who advocate banning them altogether.

But while my internal debate about the ins and outs of the ethics of zoos veers towards the 'in favour as long as they're done right' camp, I find myself wondering how I'd rationalise their existence to someone entirely unfamiliar with the workings of our planet. Humans are, as far as we know, the only beings in the history of the planet to show concern for the welfare of other species. That they mostly do so only as a reaction to the plight inflicted on those species by other humans is one of those ironies that would make it difficult to explain zoos to a visiting alien. At the heart of this irony is the parallel fact that humans are almost certainly the only species ever to understand the concept of guilt.

These penguins know no other world. They were born and raised in the temperate climate of central London, and are provided with plentiful food and a decent (if not infinite) expanse of water in which to swim. They would, like nearly all their fellow inmates, fail to survive if released. On the whole, they seem, for want of a better word, happy. But who am I to judge?

The girl presses her nose up against the glass, follows the progress of a penguin as it does laps round the pool, gives it a little wave, and says, 'Bye, penguin.' Then she scurries off to catch up with her mother.

From afar it looks like geometry, a tessellation of yellow, white and black. Only when you get closer do you see that each triangle has its own identity. They're the faces of Bewick's swans – lots of Bewick's swans – each one in three poses, like a repeating gallery of cygnine mugshots, facing left, right and forwards. There are dates and there are names. Names like Flighty, Forklift, Goblet and, more disturbing, Y-Front and Chrimble.

You see the point of it only if you look closely. This isn't Warhol for swan faces. All these beak patterns, like fingerprints, are subtly different. One, O'Hara, has a thick band of yellow – the yellow of ripening quinces or silver birch leaves on the turn – running all the way down from the forehead, constrained by the thinnest threads of black; next to it, Cattie has a Rorschach Test bill, black blotches on hi-vis yellow reminding me of a snail in silhouette, antennae standing to attention.* Others, more black than yellow, are lent a glowering look. Infinite variations, bestowing individuality on birds so uniform of plumage.

These thumbnail sketches, pages of them, are in a notebook. It was once owned by Sir Peter Scott – conservationist, artist, Olympic sailer, British skating champion, naval officer, politician, writer, television presenter, founder of one of the leading wildlife charities in

* You, no doubt, would see something different, which is kind of the point.

the world, and the kind of high achiever you instinctively want to hate, but can't because it's abundantly clear he was loved and admired by everyone who came into contact with him.*

I look at the sketches while standing, more or less, where Scott stood when presenting the television programme *Look*, which brought him and his work to a wider audience in the 1950s. It's a large room – studio, office and sitting room combined – in Scott's house in Slimbridge in Gloucestershire. Dominating it is a huge window – ten feet across and eight feet high – through which we can see a lake. And on the lake are birds. A lot of birds. I don't have a clicker with me for a more accurate count, but I'm guessing three hundred or so. There are tufted duck, pochard, shelduck, pintail, mallard, greylag, teal, wigeon, coots, moorhens, mute swans and Bewick's swans – and those are just the ones on the water. As we watch, a flock of two hundred lapwings (again, I'm guessing) swarms across the sky; a similar number of dunlin whizz round behind them – a mesmerising, undulating ensemble of striking unanimity of movement; starlings dart about busily; four carrion crows . . .

But you get the idea. There are, as I say, a lot of birds.

Beyond the lake, behind a bank, is an expanse of grassy salt marsh. And beyond the marsh is the River Severn, the Forest of Dean a looming presence in the distance. Scott described the view as 'a picture of endless beauty', and I'm not about to disagree with him.

* He was, of course, the son of Robert Falcon Scott – Scott of the Antarctic – his mother Kathleen was an eminent sculptor, and his godfather J. M. Barrie. Anything else? Oh yes, his first wife was Elizabeth Jane Howard, who went on to be a successful novelist (*The Cazalet Chronicles*). And the main protagonist in Paul Gallico's *The Snow Goose* was unloosely based on Scott. I think that's the lot. For the moment.

I'm on a guided tour of the house, so, much as I'd like to leaf through the notebook, touching it is off limits. Barbara, our guide, has been friendly but firm. No touching. No leaning. No sitting. No photography. I understand the reasoning – the house has only recently been opened to the public, and the last thing they want is unnecessary breakages. But having been told all the things I'm not allowed to do, I now have an unexpected urge to do each of them, one by one, in alphabetical order.

There are eight of us in the group, shuffling round the house, diligently steering clear of anything that might be damaged. There are designated chairs in each room for those who need a rest – not the comfy armchairs and sofas, worn and used and loved by the Scott family, nor even the 1950s window bench in the kitchen that meant they could watch the activity on the lake while eating their breakfast, but generic wooden ones offering respite to those for whom long periods of standing are problematic. One of our number – game of spirit but frail of body – takes every available opportunity to do so, balancing her sticks between her legs and allowing her body to rest while Barbara tells us about the history of the kitchen table, or the time Prince Philip came to lunch with the Scotts and ended up carving the chicken, or the provenance of the goose egg on the sideboard.

It is hard for her, this tour. I can see it in her every move, the single step up from the entrance hall to the corridor requiring care and the support of her companion, fatigue quickly setting in, pain either ever present or lurking beneath the surface. Watching her, I suddenly become self-consciously aware of my own able-bodiedness. I can go wherever I like, whenever I like. I don't have to plan ahead, look up accessibility statements, consider every part of the trip, how it will affect me, how debilitating it might all be. Like all privilege, it's the easiest thing to take for granted. I try to imagine what it would be like if any of it – sight, hearing, walking, anything – were taken away from me.

I fail.

I suspect I would become irascible,* impatient, beset by feelings of impotence, a strong awareness of all the difficulties and injustices of life. But, equally, I would like to think that were my contact with the natural world restricted in any way, I would be able to find it within myself to appreciate what was left, to experience everything available – the view over the garden from the kitchen window, the local park, whatever it might be – even more closely, to be able to savour the small things, the easily overlooked, to take some pleasure from them.

But I just don't know. You can't, unless it's forced on you, unless you have to live with it every day.

Ten yards away, a tufted duck – in the rich velvet brown of its winter plumage – floats across the lake and out of our view. Off to the left, three Bewick's swans are being harangued by a mute swan, their larger cousin. It chases them away whenever they encroach on its patch in front of the Peng observatory. This seems harsh treatment for birds that have just flown in from the Arctic tundra and are probably in need of a bit of peace and quiet, but no doubt things will settle down soon enough and the swans will see out the winter in the peace and harmony traditional to this part of the world.

Despite the room's size, any more would be a squeeze. We're not allowed near the window for fear of disturbing the birds, and the room, like the rest of the house, has been kept more or less as it

* Yes, even more so.

was when the Scotts lived in it, with chairs and sofas and desks and, most tempting, a pair of binoculars on a tripod. They're an imposing pair, suited to a naval officer. If I stepped forward a couple of paces and squatted down I could watch the Bewick's swans through them. Or I could be really naughty and pick up the little telescope on the desk, which once belonged to Thomas Bewick and which Scott was delirious to discover when it came up at auction. But despite the synchronicity this would lend my own personal journey, such things would be a breach of etiquette, trust and the rules of the guided tour. This is just as well, because left to myself I'd run amok. There are bookshelves to explore, a wall and a half of them. A lot of birding and nature books, naturally, but I also discern from my position in front of the fireplace the reassuring and familiar heft of *Grove's Dictionary of Music and Musicians* – all nine volumes of the 1954 fifth edition. There are gramophone records, too – Mozart and Sibelius, Barbirolli and Menuhin – and any number of whatnots and trinkets dotted about the place. But most of all there is art. Paintings and drawings and sketches.

Scott was self-deprecating about his art, in that particularly English way: 'I'll never be a great painter; not even a very good one.' But to my eyes the painting over the fireplace is at least very good. Four geese, flying. There is a weight to them, an idiomatic expression of their angle of bank, and an evocative quality to the light behind that places you in the scene. And if the orange of three of the geese is on the edge of luridness, a deliberate exaggeration of the effects of the setting sun on the birds' undercarriage, then that's nothing to the fact that the fourth goose is blue.

Blue is not a colour associated with geese. In fact, as many people pointed out to him, there are no blue geese. His reply was unvarying.

'Ah, but don't you wish there were?'

This combination of practicality and imagination – presumably the result of having an explorer father and sculptor mother – made Scott ideally suited for conservation. He saw the problems, and didn't hesitate to implement solutions.

And he knew his geese.

It was to Slimbridge that Scott came in 1945, fresh from distinguished war service as a naval officer, in search of the lesser white-fronted goose, a bird that at the time had been recorded just once in Britain. The skills required to separate this bird from its close relative the white-fronted goose* are considerable – a matter of being able to spot a yellow eye ring at five hundred paces, and other similar subtleties, all of them befuddling even to the initiated.

Scott's credentials in that department were impeccable. A fair chunk of his Cambridge undergraduate career had been spent shooting wildfowl from the back of a punt, and he had also made trips to the great plains of Hungary and the Caspian steppes in fruitless pursuit of red-breasted geese. Given that his main preoccupation at the time seems to have been blasting these birds out of the air, you would have got long odds against his becoming the pre-eminent conservationist of his age. His youthful relationship with wildfowl was straightforward – he saw no contradiction between his love of the living birds and wanting to kill them. It was man's primeval instinct and so on. But people change, attitudes evolve, and by the time of his 1945 visit his motives were entirely benevolent.

* In both cases, the 'front' refers not to the chestal area but to the forehead – knowledge I discovered only after several confused hours in muddy fields and bird hides wondering why the birds' 'fronts' were a sort of muddy beige streaked with black.

He saw the lesser white-fronted geese. Of course he did. He was, like his father, a man of action and achievement. But the trip had a more enduring effect. He decided there and then that Slimbridge would become not only his home, but the home of what would turn into one of the leading players in worldwide conservation. In 1946, the Severn Wildfowl Trust – now the Wildfowl and Wetlands Trust, or WWT – was formed.* As well as its primary goals of scientific study and conservation, Scott added a third: it would be open to the public. For an organisation of this type, this was a revolutionary move. Zoos were one thing, but the intention at Slimbridge was to show people the glories of wild birds alongside a collection of more exotic, captive birds. It's a formula that has proved seductive to this day.

The basis of the collection came from Scott's friend Gavin Maxwell, at that time running a shark-fishing business in Scotland, but later famous for his otter habit and resulting book, *Ring of Bright Water*. He was keen to find a taker for his collection of about fifty geese, so this was one of those rare occasions when mutual needs are satisfied with minimum fuss.

An early addition to the collection was a pair of endangered nene† geese from Hawaii. And as I leave the tour of the cottage and make my way out onto the reserve, it's a nene goose that tries to entangle itself in my flappy feet.

* One of its early supporters was George Bernard Shaw, then ninety-one, who wrote a postcard to Scott: 'I have sent in the documents and am now, I presume, a wildfowl.'

† Pronounced 'Nay-nay', and not to be confused with the identically spelled valley in the east of England where, confusingly, Scott lived before moving to Slimbridge.

Attractive bird, the nene. Not massive for a goose, with plainish buff-and-black colouring given added interest by a deeply furrowed neck. The temptation to give this fabric-like plumage a little ruffle is overwhelming, but I suspect this would be frowned on, not least by the nene itself.

When Scott took delivery of his nenes there were fewer than fifty of them in the world. Endemic to Hawaii, the bird suffered the usual depredations of any faced with new threats. Spend thousands of years with no predators, and the sudden appearance of mongooses and cats and suchlike presents quite the survival conundrum. It's fair to say the nene might now be extinct but for the success of Scott's captive breeding programme.

This one, now walking expectantly a yard or two ahead of me, has a busy feel to it. Its call – from which it gets its name – is quiet but assertive. It feels as if it's saying, 'You did know that you can buy bird seed at reception for £2.50 a bag, didn't you? DIDN'T YOU? Go back and get some NOW.' Cruel to the core of my being, I spurn its advances.

A walk through the wildfowl collection at any of the WWT's twelve reserves would be enough to endear anyone to wildfowl. Here you will find exotic species from all over the world. In fact, you won't just find them – you'll be hard pushed not to trip over them, so inured are they to the presence of humans. My first visit here was injected with a bit of twilight excitement when a pair of mute swans took it on themselves to encourage me towards the exit with hisses and clicks.

Some of the birds are endangered in the wild, and there are information boards everywhere giving details of their status, habitat and voting intentions. Others aren't endangered at all, but I'm never going to pass up an opportunity to see a goldeneye or smew up close.

The seasoned birder walks past these birds, dismissing them for the simple reason that they're captive. I'd argue that while they might have their wings clipped, rendering them unable to achieve more than the most desultory attempt at flying, they're still birds, and therefore by definition things of beauty and fascination. But I know what the birders mean. The thrill of seeing an animal in the wild is incomparable. That primeval instinct the eighteen-year-old Peter Scott was so keen on still pertains, even if we've largely swapped shotgun for binoculars or camera. Offer a birder the briefest glimpse of a bittern, skulking in the depths of the reed bed, versus a sit-down meeting with one of the tamer-than-tame eiders floating about the place doing their Frankie Howerd impersonations, and they'll pick the bittern every time.

I understand what the WWT have done here, and I appreciate it. Slimbridge is cannily laid out. There are play areas and an outdoor cafe and a small not-much-adventure playground. And if, while they're adventuring and playing, children should happen to be approached by an inquisitive mallard then so much the better. And to get to them you have to walk through the birds. It's a kind of nature education by stealth. And maybe, just maybe, they'll be inspired by this close contact with the birds to head out to one of the hides and sit freezing their nadgers off in the hope of seeing a distant stonechat. It's a tough world out there, and it's best to learn young.

I make my way through the captive collection, savouring the vivid pink of a gaggle of flamingos, communing briefly with a petulant mallard, admiring the opportunism of a bunch of starlings – wild birds profiting from the abundance of free food. I run into the woman with the sticks, now warmly ensconced under a blanket in a wheelchair, trundling slowly round the reserve, chatting to her companion over her shoulder. She recognises me, smiles, and in tacit agreement we go our separate ways before a conversation can break out.

My destination is the Peng observatory. Carpeted, centrally heated, and just yards from the same bustle of activity on the Rushy lake we witnessed from Scott's sitting room round the corner, it offers a birdwatching experience of unaccustomed comfort. Here you can sit in warmth while swans and geese and ducks disport themselves just yards away.

Or you can be diverted, as I am, by a dead pheasant, lying on its back near the window, wings splayed. Its throat is ripped open, a violent crimson gash against the softness of its plumage. A lesser black-backed gull is pecking at it, making the most of the opportunity. This little Herzogian scene of death and violence is ignored by all the other birds, which go about their business like commuters ignoring a homeless person.

Two Bewick's swans float across, taking advantage of their antagonist's temporary absence. They are notably elegant birds, slender of neck and pristine of plumage. One of them heaves itself up onto an islet a few yards away, and I'm afforded a good look at its identifying leg ring. I tilt my head to check. ZBN.

There's a large touchscreen in the corner. Swan information at your fingertips. My experience of these things doesn't give me much hope, the cumulative effect of dozens of sticky fingers most often rendering them temperamental at best.

But I'm in luck. The screen works perfectly, and you can enter a bird's ring information and find out all about it.

Z. B. N.

Bingo.

The information comes up. I read it, turn to look at the swan, look back at the screen, and finally look at the swan again, suddenly more attached to it than I could have imagined.

This swan, a male, has been given the name By Brook. He weighs 6.5 kilograms, he is eighteen years old, and in his lifetime he has travelled 90,000 miles.

Ninety. Thousand.

There are birds that travel further. I know that. The Arctic tern takes the prize – 12,000 miles or more, from north to south, then back again, every year, in pursuit of perpetual summer. And there are older birds, like the Manx shearwater that was ringed as an adult in 1957 on Bardsey Island and retrapped nearly fifty-one years later.

Eighteen is old for a Bewick's swan. The oldest recorded was twenty-four years old. And now I think of those eighteen return trips, hazardous journeys from Gloucestershire to the Russian tundra, returning to the same place each time. It's impossible not to wonder how many more he'll manage, when it will all become too much, when he'll no longer have the energy to heave his bulky body all the way there, will finally think whatever the Bewick's swan equivalent is of 'Fuck it, you know what? That's enough.'

And it's also impossible, as I look at his bill pattern, trying to memorise it so I'll remember him, not to think of By Brook as a person, almost a friend.

And that is how Peter Scott saw them. They came back every year, and he stood at his easel, painting them. And then one year he noticed the differences in beak patterns, and he and his wife and children gave them names like Y-Front and Chrimble, and now the staff at Slimbridge follow their lead and know all the birds by their facial tessellation, give them names,* and greet them like old friends on their annual return, and it is a fine and heart-warming thing.

Another swan hobbles up to join By Brook on the islet. I look at the ring. 915.

Allington. Male. One year old. Parents: By Brook and Keynell.

I sit and I look at the father and son, and I have a little moment, a silly moment of sentimentality and anthropomorphism, there on the folding seat in the empty observatory, thinking of journeys and parenthood and the unfathomable wonder of life.

Scott's last painting,† on display on an easel in his studio, was an example of what he referred to as 'SFS' (Standard Formula Scotts). Ducks and geese coming in to land on water. Clouds above, reeds below, silhouetted birds all around.

But there is one difference. In the background there are buildings. Blocks of flats, arranged in a pattern that is even now, with all its additions, recognisable as a London skyline. The idea of a full-blown nature reserve close to the centre of London might seem absurd, but this was his vision, and he had a specific site in mind: the four reservoirs at Barn Elms, in Barnes, which were about to be decommissioned.

Approaches were made, and soon he received a call from Thames Water.

* There's a distinct tendency towards *Game of Thrones* names, I notice.

† In the bottom corner, under his signature, you can just read the words 'Finishing touches Keith Shackleton September 1989'.

'Ah, Sir Peter, I understand you want to buy some of our reservoirs?'

'No. I want you to give them to me.'

He was a man to be reckoned with.

The ensuing partnership between Thames Water and the WWT saw the repurposing of the reservoirs, and the eventual opening of the London Wetland Centre, in 2000, eleven years after Scott's death.

It's to Barnes I go early every spring, to see the newly arrived sand martins; it's to Barnes I take any small child* who comes across my radar, to watch the feeding of the short-clawed otters that are the centre's pride and joy; and it's to Barnes I go in winter if I want to see a bittern.

Fifty years ago you wouldn't have seen a bittern in Barnes – nor, really, anywhere much else. They were close to extinction in the UK.† But conservation can work wonders, and the re-establishment of fens and marshes and reed beds means their population, while still small, is now relatively stable. And most winters, once the breeding season's done, a couple of them spend their downtime at Barnes.

How to see a bittern: find a place where there has been a recent sighting; sit; wait.

It's not easy. Their streaked plumage looks like a reed bed, and that's exactly the kind of habitat they like to skulk in. It is possible to sit twenty yards from one and not know it's there. And sometimes they refuse to come out. A good way of insuring against

* Some of them are small children in spirit only, but the principle remains the same.

† Having recovered, you'll remember, from actual extinction as a British breeding bird in the nineteenth century. It's a topsy-turvy business being a British bittern, but let's hope the curve is continually upward from here.

disappointment is to set low expectations. So I go into the hide resolutely prepared not to see a bittern.

The hide is empty, which, without wishing to throw shade at my fellow humans, is how I prefer it. I've had many pleasant encounters in wildlife hides. Temporary friendships have been forged, information shared. I have learned a lot. But sometimes all you want is solitude. And in that solitude, when all the elements are in alignment, a fine thing can happen.

I don't have a single word for it, but in my head it is this: the particular calmness engendered by sitting alone in a bird hide, looking out on an expanse of water on which nothing particular is happening but everything is happening – the fossicking of a snipe, the dabbling of a shoveler, the flitting of a loose group of starlings, the serene floating of a pintail, whatever it happens to be on that occasion – and not focusing on one element in particular, but letting the entirety of the scene wash over you and seep into your consciousness until the mind, at first so active and distracted, is becalmed, and your breathing slows and soon the nothing of it becomes everything and you slough off the cares and worries of everyday life and just for a bit all is well.

Sometimes you focus on the activity of a single bird – that little grebe, say, floating insouciantly, then diving with a little jump and coming up thirty seconds later, twenty yards to the left of where you thought it would. Or you regard it as a game of Where's Wally?, scouring the mudflats for the jack snipe someone saw earlier that morning. Or you just sit and allow the whole scene to wash over you, bathing in its peaceful glow.

And sometimes, just occasionally and only if you are in the right place, you might catch a glimpse of an otter, an actual wild otter, not the tame ones that come out at feeding time because they know which side their bread is buttered, but a wild one that has writhed and wriggled its way to the water's edge just to be with you.

Or that's what it feels like, anyway.

Water is, I think, an important component of this particular piece of mind-resting. Everything is 10 per cent better by water. We can theorise as to why exactly this is – whether it taps into our own composition, or whether it's to do with the calming influence of water's myriad rhythms and shapes, or any one of another thousand reasons or a combination of them all – but whatever the reasons, I regard it as fact.

No doubt there's a word in German for this feeling, engendered by this particular situation. You might call it 'mindfulness', if you're so inclined. A specific kind of mindfulness.

Birdfulness, maybe.

It doesn't always work. It's all very well wafting around the countryside saying 'Hello, birds; hello, sky' but we can't expect nature to make everything right just like that. And nor should we imagine it's a failsafe. That places too much responsibility on it, too much pressure. Sometimes your mind isn't in that place. But today it is.

Ten minutes into this visit, I have reached a state of 38 per cent birdfulness. I wonder idly if the reported bittern is around.

And now, as if conjured by the thought, it's there, morphed from the reeds themselves, treading gingerly on the ice, head forward, thick neck slung low. A most unbirdlike bird in many ways. An awkward, streaky tube of strangeness – the product of a margin doodle.

It gives me just five minutes before sloping back into the depths. Five minutes of the deepest satisfaction, during which I imagine Peter Scott's delight that his fantasy project has yielded such riches.

Birdfulness level: 100 per cent.

Another Barnes visit. A different hide.

A small gaggle of birders sits in calm contemplation of a peaceful grazing marsh. It is undemonstratively teeming with bird life.

I am one of the gaggle. My son Oliver – submitting willingly to birding indoctrination – is another.

We're enjoying the swoopings of the sand martins bringing food to their chicks in the holes in the sand bank to our right. Oliver identifies a pochard from the loose raft of ducks floating in the middle distance, and I near as dammit say out loud, 'That's my boy.'

Nothing is happening; everything is happening.

Behind us, a quiet thud, dully resonant on the hide's wooden floor.

He's sixtyish, white, ruddy-faced. On another occasion I could imagine exchanging birding pleasantries with him. But not today.

Today he is engaged in mortal combat with a telescope and its tripod. It is clearly new and clearly expensive, and he is clearly having one of those days.

It takes an act of will not to turn and stare. We scan the grazing marsh with unnatural intensity, aware of the wrestling match playing out behind us, but determinedly admiring the acrobatics of lapwings, the serene floatings of wigeon, the prehistoric legginess of herons. The man's struggle starts as background noise to this tranquil scene.

– *click thud clunk* bugger *scrape rattle clank* sodding thing will you just bloody well *shuffle clink thlunk* bollocks –

A little egret takes off, Persil-white stark against the drab marsh behind, wings languid, legs trailing.

– *smiffle clump thwack* THUDDDDDD –

An instant of silence containing all the suppressed energy of one man's frustration, its intensity signalling the strength of the outburst to come.

'FUUUUUUUUUUUCK! YOU FUCKING CUUUUUUUUUUUUUNT!'

Oliver is twelve. He has never heard those words.*

He has now.

The words ring round the hide, sinking into the bare wood. They're in the building's fabric now. Whenever I visit this spot, I will hear their echo. It's safe to say they're not exactly in the spirit engendered by Sir Peter Scott, and I imagine his brow furrowing in disappointment and consternation at their use – not particularly because of the choice of words, but because their sheer volume is likely to disturb the birds.

There is a peculiarly British scene at play here. The suppressed rage, the infuriated outburst at an inanimate object, the decibel level worthy of a Canada goose in full spate – all point to a particular form of suppressed emotion. Yet his outburst is met by a conspiracy of silence from the other occupants of the hide. There might be shock, disapproval, sympathy, a mixture of reactions. Perhaps, like me, they recognise how easily it might have been them doing the swearing.

While his mumblings were background material, it was possible to ignore him.

Not any more.

* Allow me this small fiction in the interest of storytelling.

I turn my head. He meets my eye, the simmering fury momentarily under control. By his feet, the telescope and tripod remain tragically unconjoined. I want to offer help, maybe a word of sympathy, but something about the look in his eye warns me off. He's not ready. Not yet. And he has a weapon.

It would take a hard heart not to have some sympathy; equally, it would take the nature of a saint not to stifle an internal laugh. In my case it's not a laugh of mockery – not entirely, at least. Because I know this fury, the unquenchable rage reserved for recalcitrant inanimate objects. I have felt it, and expressed it. I like to think it has calmed down over the years, and certainly it rears its head less often since the calming influence of parenthood.* But its potential remains.

I look out at the grazing marsh again. The birds go about their business as usual, unaware of the human drama afoot.

I want to beckon the man across, invite him to forget about the telescope and tripod, show him the pintail that has just flown onto the marsh, help him remember why he's here.

But then I remember I'm British, silence my inner busybody, and mind my own business. We leave the hide, and I hope against hope that his fury will be assuaged by a few minutes of birdfulness.

* My family, reading this, is no doubt stifling snorts of laughter.

8

SPECIAL JOURNEYS

In which our author honours a remarkable woman, runs the gauntlet of tern attack, and succumbs to the heady lure of the twitch

It's not going to turn up.

I've stood patiently for half an hour or so, staring at a stand of conifers in the middle distance, and the only activity to speak of was a dunnock scrabbling around on the path behind me. It looked at me with a defiant glint in its eye, as if to say, 'It's not going to turn up, you know', and then pinged off into the gorse.

It's not going to turn up.

Story of my life.

I haven't been there long, by some people's standards. A dedicated birder will think nothing of standing by a telescope for seven hours in gale-force winds, waiting for a Siberian accentor to show its face. My approach is generally to wander around a place and see what, if anything, is around.

But I'm in no hurry, the day is warm, and even without the bird it's a pleasant spot. Besides, this would be a bonus sighting rather than the object of the whole visit. I was there anyway, and when they told me at the visitor centre that there was a hobby knocking about the place, and that it had been putting on a show, it seemed a sensible thing to hang about for a bit and see what happened. Warm sun, nice scenery. What more could one want?

One could want a hobby putting on a show. Just for a bit.

Perhaps I've been spoiled. Most of my encounters with this particularly thrilling bird of prey have caught them in show-off mode, flinging themselves acrobatically across the sky in pursuit of dragonflies, their ability to turn at speed and outmanoeuvre even the nimblest of prey breathtaking to behold. Roughly the size of a kestrel, but more angular, they're a summer visitor from Africa, and very much worth the effort.

It would, in short, be nice to see one. At this stage I'd even settle for five minutes watching the bird doing absolutely nothing.

'Absolutely nothing' is a fair description of what goes on when birders gather together in anticipation of the appearance of a rarity. The pattern often goes something like this: rare bird appears in unlikely place, is seen by birder; word spreads, birders appear; bird decides it would rather skulk at the back of a hedge for the next three days, thanks very much; birders hang around anyway; bird shows its face for ten seconds, to cries of hoop-la and a barrage of shutter clicks.

There's more to it than that, of course. The few such occasions I've attended have been uniformly friendly affairs, gatherings of like minds with a penchant for the outdoors and the unfashionable and underrated ability to while away a few hours doing nothing much.

I've deliberately avoided using the word 'twitcher', the lay-person's catch-all word for anyone interested in birds. A twitcher, specifically, means someone who travels – sometimes a very long way – in search of a rarity. This hobby isn't a rarity as such – they're common enough summer visitors from Africa – and the purpose of the journey, a short detour on a journey northwards, wasn't to come and see it; but my hanging around waiting for it to appear puts this excursion firmly in the quasi-twitch category.

I've come to The Lodge, in Sandy in Bedfordshire, to pay hom-age to an institution that was a great part of my life during my formative years, and to the women who brought it into existence. In my youth, The Lodge was a place of almost mythic resonance. From there, every quarter, emerged a publication almost as eagerly anticipated as my fortnightly comics. *Bird Life*, the magazine of the Young Ornithologists' Club.

I have a copy here, from January 1972. From it I learn that my annual subscription set me* back 60p, that YOC arm badges should *not* be laundered or dry-cleaned, and that I was not the winner of the 'Name These Birds' competition held in the previous quarter.

In the section 'RSPB Reserve News' I read that 'the RSPB now owns over forty reserves'. Today that number is over two hundred, and the society's membership has grown from a hundred thousand in 1972 to over 1.1 million. How far it has come in the intervening forty-eight years, and indeed since its formation in 1891.

* OK, my parents.

The circumstances of that formation were unusual, the result of the meeting of two societies with complementary aims. The Fur, Fin and Feather Club's monthly tea gatherings in Croydon were instigated in protest at the excessive use of feathers in fashion. Meanwhile, the Society for the Protection of Birds held similar meetings in the Manchester suburb of Didsbury. It might seem an unusual concern nowadays – and was regarded as ludicrous by the general public at the time, much as the idea of banning smoking in public places would have seemed ludicrous in the 1950s – but the plumage industry was massive business at the turn of the twentieth century, worth about £20 million (approximately £204 million today), and causing untold death and suffering to the birds whose lives and feathers were taken so that society ladies could outdo each other with ever more extravagant adornments. This wasn't just a matter of picking the feathers up from the ground. The fashion reached such levels of lunacy in the 1890s that hats were commonly adorned with multiple corpses of whole birds. And even if only the plumage was required for the decoration, the birds had to die for the harvest to take place. In some cases, this mania had catastrophic effects on populations of individual species. The head feathers that give the great crested grebe its name, for example, are at their most resplendent during the breeding season, and demand for them had seen the bird driven close to extinction.

The common distinguishing feature of both organisations was that they were founded and run by women. The names of Etta Lemon, Eliza Phillips and Emily Williamson aren't as familiar as they ought to be, considering that their bold campaigning was the catalyst for what has become the largest conservation charity in Europe. In both cases, the fledgling organisations had their roots in rejection by the male establishment. The clubs and societies that might have been fertile ground for the women's ideas – the British Ornithologists' Union, Linnean Society and so on – were exclusively for men, so the only course of action for serious female activists was to form their own clubs. And the tea party, for all its genteel image, was at the heart of much political activity at the end of the nineteenth century.

In order to prevent the two organisations treading on each other's toes, a meeting was brokered by the RSPCA, an agreement was reached, and they were amalgamated, retaining the more formal name, the Society for the Protection of Birds. Royal charter was approved thirteen years later.

After the amalgamation, Emily Williamson stepped back from frontline activities, accepting the honorary role of vice president and involving herself more with local affairs as a social worker and secretary for the Didsbury branch of the RSPB. It was Phillips, and especially Lemon, who waded into the fray, campaigning and agitating and ruffling the feathers* of the male establishment.

By any standards, Etta Lemon was a force to be reckoned with. She was fuelled by a passionate love of wildlife, and especially birds, and refused to accept defeat. As a young woman she took note of all the plumage adornments worn at her local church in Blackheath and wrote personal letters to all the women responsible, admonishing them for their role in the barbaric slaughter of innocent birds. And as the activities of the (R)SPB moved into full swing and the organisation grew in size and influence, she developed a dual reputation, summed up by her two contrasting nicknames: 'The

* I am so, so sorry.

Mother of the Birds' and 'The Dragon'. So fearsome was she that a director of the Natural History Museum once hid in a stairwell to avoid her wrath, and according to pioneering birder H. G. Alexander, 'She knew exactly what she wanted the RSPB to do, and she usually got her way . . . There was no point in fighting Mrs Lemon. She would defeat you sooner or later.'

What she wanted the RSPB to do, in the early years, was enshrined in their constitution:* 'Members shall discourage the wanton destruction of birds and interest themselves generally in their protection.' It was an indication of the society's sincerity that membership also carried an obligation to 'refrain from wearing the feathers of any bird not killed for the purposes of food' – it might also have helped that, in those very early years at least, it was almost exclusively an organisation for women. The growth of the society – from one thousand to ten thousand members in just two years – was down to skilful organisation. Fifty local secretaries had an obligation to recruit members, a task they carried out with great efficiency, no doubt galvanised by Lemon's relentless cajoling.

Meanwhile Lemon herself defied the inevitable mockery that came her way, and lobbied the powerful with unyielding determination. Even though she had to give up her position as honorary secretary when the society received its royal charter in 1904 – women weren't allowed to hold such positions – she remained the driving force behind its activities for many years. It was in no small measure down to her efforts that the Plumage Bill was introduced to parliament in 1908. It took fourteen years to pass, and even then it was something of a hollow victory. While the importing of exotic

* Drawn up by Etta's barrister husband Frank.

plumage became illegal, it remained legal to sell and wear it. As she wrote in the society's annual report for 1921, 'It is impossible to say that the Act is a wholly satisfactory one.'

But these battles are slowly won. By now, the issue of avian welfare was entrenched in the public's consciousness, and even though Lemon's time at the RSPB came to an unhappy end in the 1930s – victim of a campaign led by young men who felt that the society needed a change of approach – she could point with justification to a lifetime's hard work championing the rights of birds, and the establishment of a national institution.

The aims and activities of the RSPB have come a long way since its formation, but at its heart remains an honourable ideal, and for all the inevitable faults of any organisation that has become far larger than its founders could have imagined, its contribution to the welfare of birds can't be denied.

Nowadays, the RSPB provides many people's introduction to nature in general, and birds in particular.* Unlike the Wildfowl and Wetlands Trust, there is no captive element on the RSPB's reserves – all the birds are there by choice, and they can come and go as they please. They mostly come, often in abundance.

A lot of the RSPB's members are content with passive membership – happy to 'do their bit' by joining, but limiting their activities to the annual Big Garden Birdwatch and buying the occasional blue-tit mug from the online shop. They never actually get round to visiting a reserve.

* The WWT does as well, and it would be remiss not to mention the many local wildlife trusts around the country doing exceptional work behind the scenes.

But if you do visit, you're in safe hands. If the size of the RSPB inevitably means it's unable to please everybody – Etta Lemon would be unsurprised to see that wildlife conservation remains a highly politicised area – that's not an accusation you can level at the blue-polo-shirted volunteers at their reserves, who for many people are the first and only human contact they have with the organisation.

They are cheery; they are helpful; they are informative. They will tell you where the black redstart was seen, which trail to follow if you want to catch a glimpse of the cattle egret, and the precise spot where you're most likely to see the hobby doing, or in this case not doing, its thing.

Just as I've decided to leave, I'm joined by two people, a man and a woman. He has salt-and-pepper hair, a dapper appearance, and a solicitous manner; she is frail, stooped, holding his arm with one gnarled hand. Mother and son, at a guess. We are some way from the visitor centre – quite a walk for her on uneven ground. But she gives the impression of one who gets there in the end. I wish I had better news for them.

'Any sign?'

On occasions like this there is no need to be more specific.

'Sorry, no. It should be over there somewhere.' I gesture vaguely towards the conifers. 'But it seems pretty quiet all round.'

'Pretty quiet all round' is birding code for 'there's fuck all'.

There's nowhere for the woman to sit, but she's obviously made of stern stuff, and they settle in alongside me, scanning the trees. She has a small pair of binoculars looped round her wrist, the left lens mended with tape.

Minutes pass. It is a pleasant enough scene. Calming, restful.

There is a tacit agreement between us that small talk is not required. Nothing happens, but it does it in an entirely benevolent way. A red admiral butterfly flits across above the expanse of scrub and gorse in front of us. I wonder vaguely whether, if you traced its path, it would reveal a secret message: 'HOBBYS NOT COMIN M8.'

After about twenty minutes, she turns to him. A sturdy voice, at odds with the frailty of her body.

'Shall we go back? I'm seizing up.' She gives me a look of sharp intelligence above her glasses, one eyebrow raised. 'Birds.'

We exchange farewells, briefly bonded in friendly failure, and they shuffle off round the corner, back the way they came, understandably taking the shortest route back to the car park.

I check the time. I should make a move as well. Perhaps I'll complete the longer loop, taking in some bits of woodland and views across the heath.

I hear footsteps behind me.

'Excuse me?'

I turn. It's the man. Fast now, still dapper, slightly breathless.

'It's turned up.' He makes a vague gesture behind him. 'Just seen it catch a dragonfly.'

I'm strangely touched by this small gesture, a moment of solidarity towards a complete stranger. He could have stayed there, watching the hobby. But he didn't, and I thank him for it rather more effusively than he seems to think necessary as we walk back round the corner.

The path opens out, and to the left there is a clearing. I walked past this spot forty-five minutes earlier. There wasn't a hobby here then, and there isn't one now. Just the old woman, with a wry and wicked glint in her eye. She speaks loudly and clearly.

'It's buggered off.'

Birds.

Hundreds of guillemots and razorbills stand on the cliffs, each one holding on to a tiny nesting spot against stiff competition. On the water, hundreds more, bobbing benignly in loose rafts. A guillemot launches itself off the cliff, plops onto the water and propels itself across the surface with frenetic scrabblings of its little wings. Like all members of the auk family, they're excellent divers, but not so good in the air, their wings evolved more for flipper-like underwater manoeuvrings than aerodynamic efficiency. As a result, everything about their flight screams, 'Ohgodohgodohgodohgod I'm going to crash please don't let me crash.'

And then there are the puffins, for many people the main feature. They are undeniably attractive, peeking out from their burrows, hopping up onto a rock, apparently showing off to the crowds. Our anthropomorphic nature is drawn to their quirky appearance. It's not just the distinctive kaleidoscopic bill – there's something in their facial expression that appeals to us, a clownlike, happy-sad demeanour that is entirely a figment of our imagination. Yes, their appearance catches the eye, but they're not as special as we like to think they are. If I had to pick just one species from the abundance of breeding birds on these islands, I'd choose the shag, that diminutive bottle-green cormorant, with its perky crest, steep forehead, hooked beak and hint of the prehistoric.

All these and more can be found on the Farne Islands, off the Northumberland coast, a favourite venue for anyone wanting an intimate encounter with wildlife. You get closer to the birds there than you would at any zoo, and even before the boat lands at our first stop, Staple Island, we're treated to a display of seabird activity that would count as a highlight of most trips.

'We', on this occasion, are a boatload of about thirty people. Casting my eye around my fellow passengers as we leave the harbour at Seahouses, I'm struck – not for the first time – by certain similarities. We are mostly adults, mostly men, mostly photographers. All white.

The imbalance in the demographic of nature-lovers is deeply entrenched, and while it is to some extent being addressed and corrected – not least by tireless campaigning by a few bold and committed individuals – it sometimes feels as if progress is glacial. The stereotypical image of a birder – middle-aged white male in some sort of camouflage gear, probably bearded, definitely nerdy* – might have been based on reality once upon a time, but even if it no longer completely pertains, there are strong elements of that type still in circulation. And they can, even if unintentionally, exude an intimidating vibe.†

I speak from experience, albeit of the mildest kind. When I started birding a few years ago, I felt nervous about going into hides occupied by other people. Somewhat irrationally, I felt I wasn't qualified to be in there, that I needed to know more before being admitted to the club. As I gained confidence, I realised that these inhibitions were entirely of my own creation, but if I, with all my privilege, can

* Yes, hello. That's me. In part, at least – I can't grow a beard and camouflage gear turns me off. But at least three and a half of the other four designations apply. Sorry.

† There are a very few who do it intentionally. Needless to say, they are beneath my contempt.

feel nervous, I can only imagine the levels of intimidation felt by people who spend their lives feeling excluded, for whatever reason. And for all the people I see on my visits to these places, I think of all the others who aren't there but might be, and wonder what it is that stops them from visiting.

These barriers – real and imagined – are like blueberry stains on a white shirt: absolute hell to shift. So, for all that I have seen more diversity in my journey round the country than I would have seen even five years ago, the evidence is copious that people who don't fit into certain categories feel excluded from an appreciation of nature.

There are all sorts of reasons for this, beyond the scope of this book. Perhaps they're just not interested; perhaps they don't feel it's 'for them'; perhaps they do feel it's for them, but are bullied by peer-group pressure into thinking it's uncool – it takes bravery to defy that kind of thing; perhaps there is indeed a conspiracy, however tacit, to exclude people who don't somehow 'belong'. Whatever the reasons, it needs changing. This stuff – what we call nature – should be for everybody. But it isn't. Not really. Not while these unconscious biases abide.

Right, where was I? Oh yes, shags.

While shags would be my bird of choice, there's no denying that the lure of the puffin with sand eels in its mouth is strong. And the desire to get the perfect shot can lead to behaviour more suited to Boxing Day sales than a nature reserve.

We're patient enough getting off the boat at Staple Island. The British instinct for 'no no after *you*' holds firm. But give a wildlife photographer a sniff of a puffin and the veneer wears thinner. There's no outright rudeness – courtesy is maintained at all times – but there's an intent look about them, a firmness to their resolve. They will take superb photos; I will look to spend my time communing with seabirds instead.

It doesn't take long to walk round Staple Island, but progress is slow, not just because you have to take care not to come a cropper on the uneven rocks, but also because you might trip over an eider duck. It sits stolidly on its nest, watching the steady flow of human traffic passing by within a couple of feet, but not budging an inch. These encounters are common, and it's striking how many birds are apparently unconcerned by the proximity of the human visitors. A shag sits on its nest, staring defiantly at the loose throng of people who stand behind the rope watching it; a razorbill shifts its position, revealing a single speckled egg, stretches, flaps its wings, settles again; two black-headed gull chicks, salt-and-pepper fluff, make a tentative foray from the confines of the nest, watched over by a concerned parent. There's a great black-backed gull a few yards away. It will have them on toast given half a chance.

There's no need for binoculars to study all this – the birds are within touching distance. And the humans mill around like visitors to an art gallery, looking, assessing, moving on.

While the proximity of thousands of birds brings a thrill to the heart and a whiff of ammonia to the nostrils, the activity on Staple Island is gentle compared to the frenzied attack of the Arctic terns that greets us when we land at our second stop, Inner Farne.

There's one on the path ahead of me as I trudge up from the harbour, and it is, to say the least, a noisy bugger. For the moment it restricts its protests to noisy squawks from its nest on the ground, but up ahead its colleagues are more active in their defence. They fly up, angular and aggressive, shrieking, flapping, doing everything they can to repel the intruders. Some are in your face like an argumentative drunk; some opt for diving and swooping attacks, more bluster than action, and some give you a sharp peck on the head for your troubles before turning their attention to the next in line. Protective headwear is recommended.

Just a normal spring day on Inner Farne.

I'm inclined to side with the terns. All they want is to nest in peace – they've come a long way, after all. Their migration is the longest of any bird. One individual, fitted with a tiny geolocator that used light levels to calculate latitude and longitude, left the Farne Islands on 25 July 2015, skirted France, Spain and western Africa, rounded the Cape of Good Hope, then spent time in various staging areas in the Southern Ocean before returning to the Farnes on 4 May 2016. Its total journey, in search of perpetual summer, was nearly 60,000 miles. Over a lifetime of fifteen to twenty years that adds up to about a million miles, give or take. Fly me to the moon four times, O Arctic tern.

There's one path from the landing stage up to the information centre, then a circular walk round the island. None of it takes long, but it's easy to be distracted by the terns. They nest everywhere: on the path, by the path, in the courtyard, in the entrance to the chapel. And they spend most of their time warding off potential predators. Even though the National Trust restricts the times when tourists are allowed to visit the island, this adds up to a lot of energy expended that might usefully be saved by nesting somewhere else, away from human interruption. I'm enthralled by the closeness of these birds. But I'm also uneasy. This is nature tourism at its peak. The birds do well here, but I can't help wondering if they'd do even better if we left them entirely alone. I put it to one of the volunteers.

'Yeah, it does seem odd, doesn't it? But it turns out that they use us as protection. Their natural predators, the bigger gulls, don't come near this area because of all the people, so these birds nesting here actually have a higher breeding success rate than the ones on the beach.'

Well I never.

Misgivings aside, the experience is undoubtedly thrilling. The combination of sheer abundance and close contact overwhelms the senses. And much as I appreciate the purity of observing the ebb and flow of the year's rhythms on my local patch, from time to time you yearn for something different. Allowing nature to come to you is all very well, but for some spectacles you need to travel. The day I see hundreds of Arctic terns sitting in my back garden in West Norwood will be a very strange day indeed.

The experience is made all the more extraordinary for the knowledge that these are truly wild birds apparently willing to live, for a short season at least, in close proximity to us. And if I find the existence of zoos acceptable, surely I can approve of this place, where people can have an intimate nature experience?

As I walk round, enjoying the views back to the mainland and the constant hive of activity around me, I become aware of two people whose nature experience isn't panning out quite as they'd hoped.

They are, I assume, a father and daughter. He's tall and slim, and to my eyes far too young to have what appears to be an eight-year-old daughter. What they're trying to do is get past an Arctic tern's nest on the path in front of them. The tern is having none of it, coming at the father with persistent fury. His reaction, possibly exacerbated by his lack of protective headgear, is not notable for its calmness. He flaps, he hollers, he makes a palaver omelette. Behind him, the daughter, likewise unbehatted, cowers. Her fear is understandable. The bird is intimidating, and the father makes no attempt to reassure her or help her in any way. And to be fair, this is one of the more aggressive birds we've come across.

But I'm not inclined to be fair. It's not as if we weren't warned. You can't come to the Farne Islands without being made aware of what these birds do. From the National Trust's website to the safety announcement on the boat, it's impossible to avoid the message: wear a hat.

But people don't read stuff. It's a simple fact of life.

The father makes it past the danger zone, then turns and commands his daughter to come on poppet we've got to catch the boat.

I mean, honestly.

I'm overcome by an overwhelming urge to interfere, my irritation surging up and nearly taking control. I want to say to the girl, 'Don't worry about the bird, it really isn't going to hurt you. Be brave, walk slowly and calmly, and you'll be fine. Oh, and don't listen to your father – he's an idiot.'

But wisdom prevails, and I hush my mouth. And at that moment, a photographer barges past with a muttered 'Give me a break', diverting the Arctic tern, which flurries round him for a few seconds, giving the girl her chance. She plucks up courage, makes a dash for it, and evades the attention of the tern, scurrying off to join her father. The tern throws in the towel and turns its attentions towards me. As a reward for my curmudgeonliness I get a sharp peck on the head that, even through my hat, draws blood.

Sometimes, karma works very quickly indeed.

*

The question comes out of the blue – my internal monologue, insistent like a child.

'What exactly are you doing here?'

'You know the answer to that. I've come to see the thing.'

'Yeah, but why exactly?'

God give me strength.

'It's . . .' I don't want to say 'rare'. 'Different.'

'So were the seals. And you didn't spend two hours on the riverbank waiting for them to appear. They were just there.'

'This is . . . different different.'

The appearance of the internal doubting voice isn't a surprise, in truth. I've consciously wondered what I'm doing there. An innocent morning's birding has turned into what I can only describe as a twitch. And I'm not even looking for a bird.

The scheduled part of the morning has gone well. Up betimes, heading to north Kent and the RSPB reserve at Cliffe Pools. It's a regular haunt, easily and quickly accessible from home, especially if you get up early enough to beat the traffic. On this occasion I've had the extra incentive of a harvest moon. I've watched day break from the viewpoint they call the Pinnacle, the heavy moon hanging large over the lagoons and pools and the Thames estuary beyond. Then a leisurely walk around the lagoons to the sea wall. Avocets, godwits, a bonus kingfisher. Birds, birds, birds. Lovely stuff, enhanced by the complete absence of humans and the general pleasantness of the surroundings.

From the sea wall I look across the Thames, finally picking out a dozen seals hanging around on the opposite shore. And then the idle thought at the back of my mind coalesces into action.

It was first seen the day before. I read the reports, noted that it was a couple of miles from my planned excursion, logged the information for possible use.

And then someone saw it again and curiosity got the better of me.

Beluga whales don't usually hang out in the Thames. It should be in the Arctic. That's quite a detour. But accidents happen, and now instead of socialising with its friends in the waters around Svalbard it's found itself outside Gravesend. Quite the comedown.

The decision once made, finding out where to go is simplicity itself, thanks to the active online network that shares news of unusual sightings.

I drive, park, walk, assuming that somewhere along the banks of the river there will be a gaggle of twitchers, optical equipment pointed towards the murky waters.

As ordered.

I come across a man, his face as long as the massive telephoto lens on his camera. Ten yards down the path, a group of four. And about twenty yards further down, a disconsolate looking pair of middle-aged men with binoculars.

I might have come to the right place.

I stop by Massive Lens Guy.

'Anything?'

'Nah.'

'It has been seen today, though?'

'Yeah.'

'Thanks.'

He's not unfriendly. Just bored. There's something about his demeanour that speaks of an unwillingness to be here, which makes me wonder why he is, in fact, here. I try my luck.

'You a whale watcher in general?'

He lights a cigarette and throws me a disdainful look.

'Photographer. Paper says go, I go.'

'Ah. I see.' And I do. The whale was first seen late the previous afternoon by a sharp-eyed birder. Word spreads quickly, and soon enough news outlets had picked up on it. A beluga in the Thames is unusual enough to be newsworthy. Photographs and video footage, for a while at least, have currency.

After a couple of minutes he throws his cigarette away and seems to decide to give me the benefit of the doubt.

'How about you? Whale watcher?'

'Birder.'

An upward nod, soft grunt.

'Come far?'

'I was kind of in the area anyway, just down the road.'

Another nod. Then he holds his hand out.

'Steve.'

'Lev.'

It feels somehow old-fashioned.

It's a nice enough morning, without being spectacular. Thin high cloud stops the sun from working its magic, and the Thames looks an unappealing prospect, brown and murky. It's warm enough to be able to hang around without ossifying, but not balmy.

Time passes. The beluga stays resolutely out of sight.

'Course, it might not appear this side.' He gestures across the river, about a kilometre away. 'Might choose to go Essex side.'

I raise my binoculars. There's a similar-sized gaggle of people on the far bank.

'Is that where it was yesterday?'

'Nah. It was this side. We're OK.'

More time passes. The surface of the Thames remains untroubled by cetaceans.

A cyclist, lean and Lycra-clad, trundles towards us.

'You looking for the beluga?'

We nod. He points to where he's just come from.

'Up there, about two hundred yards. Been doing its thing for half an hour now.'

I blame myself. I'd seen Steve and assumed he knew this was the best spot. Rule One: never assume anything. The other guy doesn't always know best. I've had some of my best sightings by looking in the opposite direction to the one chosen by the serried ranks of telescopes.

We walk along the river, and sure enough, there's another group of six people lined up on the bank. We join them and wait.

It doesn't take long. And naturally I miss it. All I hear is the chattering of camera shutters. *Tchrrrrrrrrrtch*. And by the time I've looked, it's gone. But I don't have long to wait. I scan the area, unblinking. After a few seconds, the murky water is breached again. The merest glimpse. A milky parcel, rolling through the water, and then gone. There are two more sightings, frustrating in their brevity, and then it goes quiet, and we're left looking at the murky water and wondering when it will appear again.

Behind us, meadow pipits, dunnocks, starlings and goldfinches provide an everyday display of brilliance, if only we could be bothered to turn round. But we are, temporarily at least, immune to the seemingly mundane.

Overhead, the drone of a helicopter disrupts the peace of the morning. And now a boat appears, two figures at the front, long lenses clearly visible from where we're standing.

Slow news day.

The boat's proximity to where the whale was last seen causes agitation in the ranks of the twitchers.

'That's too close! Too sodding close!'

'Idiots.'

Steve smiles quietly.

'It's OK for some, isn't it?'

'Know them, do you?'

'Oh yes.'

'What paper?'

He tells me. I'm not surprised. The sun chooses this moment to appear from behind a cloud.

'To be fair, they're just doing their job. I'd do the same. I wouldn't like it, but I'd do it. Work is work.'

'They're not overly concerned with the welfare of the whale, then?'

He flicks me an amused look.

'Nah, they couldn't give a shit.'

Somehow the message gets through to the boat that its presence is frowned on, and it moves upstream. But its work is done. The whale might be hunkered down a little lower in the water, waiting for the fuss to die down; it might be making its way by stealth towards the Essex side; it might have found its bearings and started the long journey back to the Arctic.

I stick around for a bit, but soon the sun goes in and there's a hint of drizzle in the air. I'm done. I've seen it, ticked it off.

I'm simultaneously moved and underwhelmed by the experience, and not quite sure whether I feel better for it. On the one hand, it was just a fleeting glimpse of a pale white tube breaching the surface of the water; on the other, it was an insight into the life of an alien being, prompting thoughts of vulnerability and strength and perseverance, but undercut with the knowledge that in ideal circumstances I wouldn't have seen it at all.

And I can't shift the memory of the shameless, nameless photographers on the boat. The contrast between their lack of regard for the beluga's welfare and Etta Lemon's utter dedication to the idea that nature requires and deserves our help couldn't be starker. And the realisation that 'couldn't give a shit' sums up many people's attitude towards nature – more than I care to think about – is profoundly depressing.

But then the memory of a forgotten moment, triggered by the vulnerable cetacean I've just seen briefly in the murky waters of the Thames, surges up. A moment on the journey back from the Farne Islands, when the surface of the water fifty yards away was breached by one, two, three dorsal fins, and someone shouted, 'Dolphins!', and everyone turned and looked, and there they were, five of them now, a pod of bottlenose dolphins, their progress fast and clean, their rhythm and flow somehow expressing a freedom to which we all aspire.

Dolphins are like that. We see them moving through the water, picture their smiling faces, and we get all anthropomorphic, imagining them carefree and joyous, the embodiment of the glories of nature. And as I watched, suddenly smiling, a man next to me helped his small son stand up on the seat, gave him his binoculars, and held him steady while he watched with an expression of breathless excitement and wonder.

A moment to give you hope, to remind you that in the vast gulf between the two extremes of engagement there is more good than ill, that if you chip away, show enough people the magic of a dolphin or a hobby or even the dunnock on the path, the tide will, against all the odds, make a turn for the better.

9

WILD THINGS

In which our author seeks wildness, pays homage to an otter and his Maxwell, and finds joy in rain

We have come to a magical place, a place of wonder. A place of eagles, seals and otters; of heather, rowan and silver birch; of mist, lochs and sheep; of serrated ridges, looming hills and sweeping, U-shaped valleys; of six different types of cloud in one sky and four seasons in a day; of waterproofs, backpacks and walking boots; of sun, drizzle and downpours; of shafts of sunlight illuminating the ageless hills in all their majestic grandeur, heads in the clouds and thin ribbon waterfalls tickling their midriffs.

A place, as it turns out, of countless camper vans.

We have come to Skye.

The city dweller in me yearns for this kind of space. A population density of six people per square kilometre should ensure that encounters with other humans are few and far between. But the road to our rented cottage is single track, and we learn quickly enough that part of our daily routine will involve a fair amount of waiting in passing places while a clutch of vehicles – resembling nothing more than massive, evil, sentient freezers – trundles stolidly towards us, their drivers' attention divided between keeping their behemoth of a vehicle on the road and glancing up at the relentlessly observable scenery.

This enforced slowing down does at least enable us to strike up friendships with several sheep, whose ability to add a level of picturesque rural charm to any tableau is matched only by their ability to trot out in front of cars at just the wrong moment.

What with the sheep and the camper vans and ooh look at the way the sun brings out the purple in that heather, progress around the island* takes a while. But there are worse places for a slow drive than Skye. It is a place of enduring ruggedness and beauty. If my adage that everything is 10 per cent better next to water holds true, you can add at least another 5 per cent if the water is fringed by hills whose immediate, in-your-face glamour gives way on closer inspection to subtler charms. Their true character is elusive, revealing itself piecemeal, and shifting according to the prevailing weather of that particular minute.

It would, in fact, be perfectly easy to spend our week on the island contentedly pootling around in the car, enjoying its delights at one remove, cocooned from the elements and experiencing the whole thing as if it were a muffled virtual reality travelogue.

* Much bigger, I soon realise, than I'd imagined it – it's sixty miles from tip to tip, and its individual shape means it has four hundred miles of coastline.

But we are a family of at least moderate activity. While I don't cycle up a hill with quite the relish exhibited by my son, and certainly don't regard a mountain as an opportunity to grab my hard hat, ropes and crampons and clamber up it like a deranged ibex, I am at least a disciple of the 'don the sturdy walking boots and see you at the top for a slice of Kendal mint cake and a wee dram' school of outdoor experiencer.

Besides, it would feel like a betrayal of the island to flap around it at the edges. You can't get to know a place without smelling its air, examining its landscape, experiencing its weather – and Skye has all three of those commodities in abundance. So we go for walks of varying lengths and strenuousness, ensuring each time that we are suitably equipped with water-excluding bodywear and a cheerful 'what the hell, it's only water' demeanour. In the event, they're barely needed. Skye's weather, much maligned, treats us very well indeed.

It all depends, I suppose, on your definition of 'very well indeed'.

My relationship with early mornings is mixed. On the one hand, there's that heady feeling of being a conspirator in the hatching of a new day, the knowledge that whatever the ensuing hours may hold, I'm giving it my best shot and not missing a minute of the action; on the other hand, my eyes are glued shut and my brain mutters darkly at me for ruining a perfectly good dream. But victory goes to the part of me that wants to explore this new and intriguing place. It always does.

The air is fresh and clean, a positive invitation to breathe deeply and exhale with a satisfied 'aaahh'. The hills in the distance are dark outlines, their silhouettes crisp in the early-morning light, detail obscured in shadow – they're holding their secrets close to their chest just for the moment. In the foreground, the loch is a dull pewter, the ripples in its surface touched by the metallic peach reflection of the rising sun; off to the right heavy grey clouds foretell wetness, leeching their 4B pencil shading down onto the surface of another, more distant loch. Bubble of twite, chatter of goldfinch, caw of hooded crow. Over the hill behind me a buzzard mews.

Toto, I've a feeling we're not in West Norwood any more.

Aware of the dangers of the instinctive urge to photograph everything in sight, I've been more reticent with the camera of late, but this scene is so blatantly snappable that I can't resist. Besides, it will act as a perfect summation of the Skye experience, encapsulating as it does so many aspects of it. Loch, hills, sun, rain, a couple of islets poking their torsos above the surface – all in a picturesque tableau that positively begs to be photographed in the panoramic style that seems such a good idea at the time but never really comes off when transferred to the flat medium of your computer screen.

The walk from the front door of the cottage takes me to the end of the road, past the ruins of an abandoned village and beyond, onto a soggy path leading gently inland while maintaining extensive views of the surrounding ooh-ness. It's a gentle, undemanding climb and I see no reason to curtail my walk. I look back across the loch. The clouds have, if anything, intensified, an effect explained by simple geography: earlier they were far away,* but now they are close. I estimate it'll be half an hour before I'm subjected to a drenching.

Such easy confidence.

* Not small.

On the way back, I notice the effect the incoming clouds have had on the landscape. Shadows have shifted, silhouettes darkened and softened. The berries on the lone rowan tree near the ruined village, so bright before, have taken on a dull lustre. The sun, which ten minutes earlier had brought to mind such words as 'pellucid' and 'luminescent', is about to be engulfed by a moving wall of cloud, bringing to mind such words as 'fuck me, I'm about to get soaked'.

I quicken my pace. Lulled into a false sense of security, I've left the house wearing nothing more protective than a pair of binoculars.

I'm stopped by a *gronk*. Or maybe it's a *grawk*. However I might choose to transliterate it, it's enough to stop me in my tracks and bring my binoculars to my eyes.

This is not a sound I know.

I find them quickly enough. They're flying highish over the loch, heading towards me. Almost duck-like, but not quite. Almost goose-like, but not quite. A hint of cormorant.

Red-throated divers.

I've seen these birds just once before. Then, they were flying low above the horizon, at least two miles away over a turbulent sea off the Isle of Sheppey on the coldest day I have ever experienced.* I saw them that day only because my brilliant friend David, who can pick a needle-tailed swift from an avian haystack with his binoculars tied behind his back, showed them to me through his telescope. The

* Not to brag or anything, but I've been in temperatures of –20°C or so during a frigid Toronto winter. The Isle of Sheppey, on that windy December afternoon, probably registered only about –1°C on the thermometer, but once you throw in the Brit-chill factor of –1 gazillion, had Toronto beaten hands down.

memory of their shape, distinct even at that distance, and his pithy explanation of how he identified them – dumpy body, low-slung head – have stayed with me across the intervening three years, and now, at last, I have an opportunity to put that knowledge to good use.

They have a strange clumsy elegance about them, like someone speaking perfect French with a thick English accent. Fast but somehow awkward. And their oddness is reinforced, as they fly past and veer away from me over the loch, with another *grawk*.

Although I'm not familiar with the sounds of red-throated divers, I can tell you other things about them. Their scientific name, for example: *Gavia stellata*. Or the pleasing fact that, while we Brits resolutely adhere to the name 'diver' for this family of very much water-based birds,* the rest of the English-speaking world goes with 'loon'.

I'm fairly sure of my identification, but the last thing you want is to be the kind of birder who reels off a series of misidentifications with misplaced confidence, based on little more than a brief glimpse and a desire to have seen the bird in question. Much better to err on the side of caution. Luckily, in this modern world of immediate gratification, I have in my pocket a device loaded with an application that has an encyclopedic collection of bird sounds, and it is the work of a few seconds to confirm that the *gronk*† I heard was indeed the work of the bird they call the red-throated diver or loon.

The thrill I get from this sighting isn't to do with rarity. Not entirely, at least. A sighting of an unfamiliar bird always has its own

* They have short legs, positioned very far back on their body, and are comically ungainly on dry land.

† Or *grawk* – the jury's still out.

frisson, but the frisson alone can't account for the particular quality these birds bring to the morning.

It takes me a few seconds to work it out.

It's not just the birds, for all their unfamiliarity and weirdness. And it's not just the hills, the loch and the sky, attractive though they are. It's the whole package – all of the above, all tied up with my solitude and the strangely intimate moment with the red-throated divers into a neat little bundle of special. I can see it now. I would paint it, if I could.

And while it's quite possible there's someone a few hundred yards down the road sharing the sighting without my knowledge, I selfishly claim this moment for myself. They're my birds, my hills, my looming, massive rainclouds, now overhead and heavily, ominously pregnant.

Oh dear.

It doesn't stand on ceremony, this rain. None of that mizzle nonsense. Its opening gambit is a thick, heavy drop on the road in front of me, the kind that might ricochet and take out a small child. And immediately it's backed up by a battalion of its compadres. This is proper, serious, professional rain. Rain that would have no qualms about breaking into your house, raiding the fridge, turfing the cat off the sofa and chatting up your girlfriend. It's taking no prisoners.

I'm still at least half a mile from the cottage, and it's not going to stop to let me get back.

With sinking heart and rapidly moistening clothes, I trudge onwards. The old argument of whether it's best to walk quickly or slowly in a downpour is moot – it's going to get me, whatever.

And then I stop.

This rain isn't cold. I can change my clothes at home. I was going to have a shower anyway. What is it about the rain, I ask myself, that makes me want to avoid it? A moment or two's thought, accompanied only by the sound of litres of water hitting tarmac, gives me the answer: nothing.

That flick of the switch. The instant change from rejection to acceptance, from grumpiness to cheer, from 'no' to 'yes'. Smiling, laughing almost, I submit to it, and suddenly the rain is my friend. There's a freedom to it, a release. It's like that moment at the end of *The Shawshank Redemption*, but without the preceding crawl through miles of raw sewage.

The water cascades through my hair, penetrates my clothing, saturates me from stem to stern. On another occasion – many other occasions – this soaking would cause misery, a drowning of the spirit. But just this once it feels liberating. Perhaps it's the knowledge that dryness and warmth are not far away; perhaps the intoxication of the first morning of the trip has immunised me to misery; perhaps it's the divers.

Yes, that'll be it. Birds are responsible for everything good. Unless you can come up with a more plausible explanation.

There's a kerfuffle at the car park – on Skye, there often is. Two cars ahead of us wait for a camper van to emerge from its parking place by the side of the road. The camper van is hampered by a pothole, the proximity of a bank by the side of the road, and its driver's incompetence. The front car is waiting for the space between the camper van to be big enough for it to slide in, like a ferret sizing up a particularly tight drainpipe.

There is manoeuvring, guiding, fraying of tempers. Each thinks the other wrong, is unwilling to yield, can't believe other people don't just do it their way and be done with it.

If you want metaphors for modern life, go to a car park on Skye.

From my vantage point at the end of the queue, I can see it all playing out in ultra slo-mo, each player's move painfully obvious. Fiat Punto to camper van six, checkmate.

I am, as always, the model of patience, sitting quietly behind the wheel and taking the opportunity to admire the particular shade of golden-brown in the bracken* brushing the side of the car. Meanwhile my brain forms a silent scream.

Just another day on the peaceful roads of Skye. The hills look on from above and shrug their shoulders. They've seen it all before.

If Skye's large open spaces make its population density low compared with, say, London, then the agglomeration of people at the various pinch points causes traumatic feelings of claustrophobia. It's a strange contradiction, and a relatively recent one in the island's long history. While the influx of tourists is good for the island's economy, there's a perverse conundrum at play here. Yesterday's unspoiled idyll is tomorrow's tourist trap, and there must surely come a point when the pain outweighs the pleasure. We're all suckers, fooled by the guidebooks and TripAdvisor into visiting the 'top places'. But when you visit a beach once revered for its seclusion and privacy, and find it host to three cafes, two gift shacks and an ice-cream van, you begin to question the wisdom of crowds.

* A non-native invasive species, it is, understandably, much frowned on by the locals because of its tendency to take over, but there's much to admire in its drooping fronds for the fractal fan.

Naturally we regard ourselves, by long convention, not as tourists but as 'visitors'. It's a fine distinction, illustrated by the irregular declension: 'We are visitors; you are tourists; they are thoughtless hooligans desecrating a holy site.'

And so it is when we visit the coral beach, a place of unusual beauty and a nightmare of a car park.

The coral beach is across the loch from where we're staying, no more than two miles as the red-throated diver flies, but a forty-five-minute car journey even before you factor in the passing places. The walk from the car park is just long enough to make you feel as if you've had a walk, but not so long as to become tiresome to the non-walker. I tarry a while in the car park, thanks to a flurry of house martins overhead. I try to count them, and reach a total of forty-three before having to admit that I've counted some of them twice and settling on a provisional total of 'twenty-five or thirty, give or take'. Regardless of their number, their excitable chattering and nimble dartings are an instant smile generator. They even distract me from the midges, which choose this moment – after a surprisingly reticent couple of days – to descend en masse.

Skye's midges are famous. Any mention of the place elicits two responses: 'Ooh lovely', followed by 'Watch out for the midges!' A select minority will then offer their favoured antidote, and it is with the most popular of these that we now slather ourselves.

It works. Until it doesn't.

I decide that mind over matter is the best tactic.

This works. Until it doesn't.

And then, no more than a couple of minutes' walk out of the car park, they disappear. We barely see another one all week, a development welcomed by all.

As we approach the coral beach, we begin to see what the fuss is about. On a grey day, it has a quality of soft brightness that illuminates the area around it even from a distance, offset by water the kind of colour normally associated with the Caribbean. Cobalt, cerulean, topaz, turquoise – call it what you will, it draws you in and soothes the soul, an effect only enhanced by the hypnotic accompanying sloshing and lapping sounds.

As I said, 10 per cent better.

Adding to the beneficial effect is the grass, which butts up against the beach like carpet to a skirting board. It's soft grass, too, not the sharp kind that lashes your bare legs as you clamber over the dunes, sinking into the loose sand up to your knees. And it is speckled with flowers of yellow and purple and white, the names of which I vow to look up but still haven't. Stitchwort is one – that much I know – and devil's bit scabious another, and maybe autumn hawkbit (agh help I'm really not sure), but try as I might, I can't reconcile the little purple orchidy thing with any of the photos in the book so I'm just going to call it little purple orchidy thing and be done with it.

Never mind. They're pretty, and give the deep green grass the air of a medieval tapestry. And just thinking of the names of wild flowers gives me a bit of a perk-up – scabious, twayblade, corncockle, saxifrage and many more. Words that feel steeped in history and connection with the land.

We walk down until we feel the scrunch underfoot. It's not sand, but a fossilised and bleached coralline seaweed known as maërl.* Lime-rich, it was long extracted for use as fertiliser – from an environmental point of view it is a Good Thing, harbouring biodiversity

* Pronounced 'marl'.

from sea urchins to seals. It is also fascinating to look at. We get down on our knees to examine its varied shapes, sizes and shades: bone, ivory, off-white and all the rest of them, bringing to mind a clothes rail full of linen jackets.

There's a grassy mound just off the beach, a short effort to climb. We do so for no better reason than that it's there, but the view from the top is worth the effort. From here you can see the clarity of the water, its quick transition from lightest blue to gunmetal grey as the shore falls away into the loch.

From here, too, we can see across the water to the north, to the island of Isay and beyond to the Outer Hebrides. This beach, the furthest point you can reach on this bit of Skye, feels at the edge of something. Not the world, nothing as dramatic as that, but at least this part of it.

I have a sudden yearning to launch out, somehow, towards the open sea, to leave behind the comfort and ease and warmth and to head for somewhere truly wild. St Kilda, maybe, with its subspecies of wren that sings more loudly than the mainland ones to compete with the wind; or Rockall, perhaps, there to surprise a gannet or two.

It doesn't last. True wilderness is unbearably harsh and impossible. And the overwhelming majority of us are too softened by what we like to call civilisation. We like the idea of it, which is one of the reasons places like Skye are so popular. And the more we lose contact with nature, the more we're drawn to spend our time doing things that aim to emulate wildness – glamping, hiking weekends, cycling tours. But make us live in it and the reality would bite, and bite hard.

Back on the beach, a small child squats low, working with utter concentration. She is making patterns in the maërl, a network of spirals and circles and free loops, each line going where her fancy takes it. The lines intertwine and fold back on each other in a most peaceful way, and the result is a mesmerising improvised doodle – a Zen garden, with added freedom.

To judge by the extent of it, she's been at it for quite some time. Her father looks on with what I like to think is approval but might equally be exasperated boredom. We cannot know the workings of other people's thoughts.

We walk slowly back. On the shore, a herring gull eviscerates a flatfish. Clouds clear slightly, allowing an end-of-day sun a brief moment of glory. The house martins have gone. So have most of the cars. But the midges are still there. As we reach the car park, a boy – sixteen, seventeen – fiddles with an elaborate arrangement of scarves and sweaters, making absolutely certain that every square inch of his skin is protected.

Some people will do anything to keep nature at bay; others will allow it all the way in.

In the autumn of 1957, a young man visited a terraced house in the furthest reaches of Chelsea. He rang the bell and waited. Receiving no reply, he retreated to the pavement and tried to peer through the front window. He found himself staring at a large stuffed lizard. It was only when the lizard's tongue flicked out and caught a grasshopper that he realised it was very much alive. The brief appearance, a few seconds later, of a large and brightly coloured bird at the window gave further indication that the house was indeed occupied, and sure enough, after another ring of the bell, the door opened and a lean man appeared. He explained that he had heard the bell, but had spent the intervening minutes observing his visitor through binoculars from the other end of the room.

Had I been the visitor, I suspect I might at this point have backed cautiously away, made my apologies, and left. A monitor lizard is one thing, and to meet one face to face would be an intriguing and potentially attractive prospect; a scarlet tanager, for a young bird-lover, would exert an even stronger hold; but the revelation that my host had spied on me would have been enough to set alarm bells ringing.

The young man was made of sterner stuff. His host invited him in and poured him a half pint of whisky. Then, without saying a word, he took a small, ivory-handled pistol from a drawer, held it to the young man's head, and pulled the trigger.

And it is here that, after a due pause for hysterical blubbing and incoherent shouting, I would undoubtedly have sprinted for the door, never to return. But the young man, undisturbed by his host's patently psychopathic behaviour, stayed, becoming the older man's friend and biographer. His name was Douglas Botting; his host, forty-three at the time but with the wrinkled and lined face of one much older, was Gavin Maxwell.

You might easily conclude from the above story, told by Botting at the beginning of his biography of Maxwell, that the multitalented but chaotic aristocrat was, to put it mildly, 'a bit of a character'. And you might be right. But naturally there is a great deal more to it than that.

Maxwell was an adventurer. He went on an expedition to the Arctic tundra in search of the elusive Steller's eider; he was a secret agent in the war; he ran a shark-hunting business (it failed); he dabbled in motor racing. And even when engaged in relatively normal pursuits, drama followed him around like a cat begging for food.

He was the kind of person who could go out fishing and come back with a flightless fulmar and a stricken Manx shearwater (they lived, briefly, in his bathroom). The chaos of his personal life encompassed numerous car crashes, two fires, and multiple career changes – often inspired by wildly impractical ideas – but it was as a writer that he became known.

The phenomenal success of his fifth book, *Ring of Bright Water*, catapulted him into the public's consciousness. Published in 1960, it's the story of Maxwell's relationship with an Iraqi marsh otter called Mijbil, or Mij, given to him by writer and explorer Wilfred Thesiger on a trip to the marshes of the Tigris Basin. He brought Mij back to Britain (a journey not without its dramas – loose otters and passenger aircraft aren't necessarily a good mix) and they lived together in Maxwell's remote cottage at Sandaig* in the West Highlands.

Maxwell was, in his own words, 'otterly enthralled' by Mij. He had recently lost a much loved dog, Jonnie, but described his relationship with the otter as more one of co-existence than of ownership.

If a love story between a man and an otter seems an unlikely bestseller, *Ring of Bright Water* also owes its success to Maxwell's portrayal of the landscape of the West Highlands. It was a wild, remote and romantic world, an escape from the drab reality of post-war austerity. Maxwell's day-to-day existence in his house at Sandaig – on the mainland but looking across to Isleornsay on Skye – must have felt impossibly dramatic to the average reader. There were no amenities, a visit to his nearest neighbours meant a long slog across peat

* In an effort disguise its true location, he renamed it 'Camusfeàrna' (Bay of Alders) for *Ring of Bright Water*.

moors, Kyle of Lochalsh was forty miles away by road, and supplies came by boat and had to be carried to the house by rucksack down a long track.

Ring of Bright Water is not a scientific book – but with its whole-hearted adoration of Mijbil (and, after Mij was tragically clubbed to death by a roadmender, his successor Edal), Maxwell did more for the status of the whiskery mustelids than any amount of earnest lecturing could have. At the heart of it is his love for them, and their apparent reciprocation. He tried other pets after Mij died. They didn't last long. Kiko, a ring-tailed lemur, slashed his legs, leading to enormous loss of blood and an illustration of Maxwell's cavalier attitude to personal safety – he watched dispassionately as blood spurted from an artery while he tried to remember where the pressure point was.

No such excitements were attached to the nameless bush baby that followed, but therein, perhaps, lay the problem. He found the animal boring, and replaced it with a small flock of tropical birds.

No, it was otters that did it for him. He describes with relish Mij's morning routine: sleeping in his bed, waking at 8.20 prompt, nuzzling his face to wake him and then stripping the bedclothes to ensure that he got up. Ball games and bathtime were sources of endless entertainment for human and otter alike. And there is a carefree joy about the descriptions of Mij swimming and diving and tumbling in the sea, burn and waterfall around the house, and their daily walks along the beach.

This vivid description of a man's relationship with an animal will have struck a chord with any pet owner – that it was an otter added a touch of the exotic and eccentric to the story. And if the average person, used to no more pet-related inconvenience than fur-covered cushions or the odd bit of vomit on the floor, might throw their hands up in horror at the thought of a domesticated otter having the run of the house, it's clear that Maxwell relished it.

The connection between his love of animals and the difficulty of his relationships with humans is easy to make, and even the most amateur of psychologists will leap to link all this to his childhood. His father died when Gavin was just three months old. His mother was a rather austere woman, and his early years were characterised by the appearance of a succession of huntin' and shootin' aunts – all tweed, brogues and pipes. Nature seemed preferable.

But equally fascinating is the evolution of his relationship with nature. Like Peter Scott, he went from avid hunter in his university days to equally committed conservationist; and also like Scott, he saw trouble ahead, writing this in the preface to *Ring of Bright Water*:

> I am quite certain in my own mind that man has suffered in his separation from the soil and the other living creatures of the world. The evolution, as it were, of his intellect has outrun his needs as an animal, and now, still, he must, for his own security, look long at some part of the earth as it was before he started messing about with it.

Nevertheless, there is also a strange contradiction at play here. On the one hand, his love of wild animals; on the other, his eccentric desire to keep them captive as pets.

His last project was to be a sort of menagerie on Eilean Bàn,* a six-acre island between Skye and Kyle of Lochalsh on the mainland.

* 'White Island' in Scottish Gaelic.

He had bought the two lighthouse cottages on the island in 1963, after the automation of the lighthouse, and moved there when his home at Sandaig burned down in 1968.

He had grand plans for Eilean Bàn. It would be an otter sanctuary, but there would also be an eider colony around nearby islets, as well as a sort of menagerie, for which he had already bought a gannet, a tawny owl and some goats.

But before he could realise the project, years of ill health caught up with him – eighty cigarettes plus a bottle of whisky a day doesn't make for the healthiest of lifestyles – and he died in 1969, less than a year after he moved there.

It doesn't take a leap of imagination to guess what Maxwell would have thought of the road bridge now linking Skye to the mainland, but he might also quietly have appreciated the irony that Eilean Bàn is now the home to one of the pillars supporting the bridge – opened in 1995, twenty-six years after his death at the age of fifty-five. And as a keen and somewhat reckless racing driver he would no doubt enjoy the manoeuvre required to park outside Eilean Bàn when approaching it from Skye-side – a hurried U-turn to take advantage of a brief lull in the steady traffic.

We park in the small layby, go through the gate and down the path, and I try to imagine the house before the bridge, without the looming presence of the monumental pillars, without the rumble of cars, without the intrusion of modernity. It's an enticing picture, full of peace. David, our guide, gives us a brief biography of Maxwell, brings him to life with a couple of well-chosen anecdotes. Then he shows us Maxwell's long living room, furnished with taste, style and not a little expense. A large gilt mirror, Moroccan daggers, a stuffed jay. The telescope through which he watched boats and dolphins and gulls and ravens and sometimes a killer whale, but also the comings and goings of people on Kyleakin High Street.

We go up the lighthouse, with its views across Inner Sound to the islands of Longay and Pabay, and, more prosaically, the fish farm on the Skye coast. We spend time in the wildlife hide, from which we do not see black guillemot, shag or white-tailed eagles, but we do see a great black-backed gull, pecking furiously at a dead something on the shore.

And as we thank David and walk slowly back to the car, my eye is caught by a small and exquisitely beautiful thing. It's nothing, really. Just another rock, a smooth one with a little craggy bit at the top that would make it just too uncomfortable to sit on. But its flat surface is coated with lichen in a pattern of such delicacy as to demand closer examination. It looks like a network of misshapen fields photographed from way up high – irregular patterns in infinite variations of beige, marbled through with the thinnest lines of dark green. I take a moment to consider the strange fascination of these organisms – composites of fungi and algae or cyanobacteria, living symbiotically, manifesting in thousands of different forms. It's easy to dismiss them, but as they are estimated to cover 6 per cent of the world's surface, perhaps we should give them more attention. Sometimes it's the smallest things, the insignificant or easily overlooked, that catch your eye.

And sometimes, by contrast, they're so big you can't ignore them.

There's a sort of grim inevitability about it. The morning dawned dry and sunny, with just a few light-and-fluffies shrouding the hilltops;

the drive across the island was notable for the quality of light, the bright sunshine bringing out the best of the various shades of green and purple in the grass-and-heather duvet draped over the hills; the forecast is set fair. So, sure enough, the moment we set foot on the boat it starts chucking it down. You might take me for some sort of rain god.

This manifestation of the island's multiple microclimates is received with resigned good humour by all the boat's passengers. There are twelve of us, plus two guides. They will steer us out of Portree harbour, out into the Sound of Raasay, hugging the coastline until the purpose of the trip is fulfilled. There's a barely suppressed excitement aboard – we're off to see the eagles.

The allure of the white-tailed eagle (*Haliaeetus albicilla*) isn't hard to fathom. They're massive – nearly a metre tall and with a wingspan of up to 240 centimetres – and if there's anything we like more than things that are small and cute, it's things that are large and terrifying. The white-tailed eagle falls firmly into the latter category, sharing equal top spot for Britain's largest bird of prey with its cousin the golden eagle (*Aquila chrysaetos*). A sighting of either bird is a memorable occasion, and the white-tailed's rarity* only adds to the romance.

Sightings, I have read, are possible anywhere on the island, but we're keen to make sure, so the boat trip, with its bet-hedging assurance that 99 per cent of excursions yield a sighting, seems a worthwhile investment.

* Not that the golden eagle is common – far from it – but its range is somewhat larger, extending across a lot of northern Scotland, while the white-tailed is restricted to a few islands in the west.

Nigel, our guide and pilot, treads the fine line between jovial banter and irritating chumminess with the assurance of one experienced in his field. He dispatches the safety announcement with good humour, points us towards laminated information sheets and tells us what we can expect to see. We duly scan the water for the hint of a porpoise, with predictable results.

Our travelling companions on this two-hour excursion are, for the most part, quietly genial, keeping to themselves and speaking when spoken to.

For the most part.

There is one exception. She is talkative, she is friendly, she is completely lacking in self-awareness.

It's clear from the moment we board that we're in for a running commentary. As she steps from harbour to boat, she nearly comes a cropper.

'What am I like? I nearly fell in there!'

Unable to think of anything to say, I say nothing, hoping that a non-committal quarter-smile will speak volumes.

No such luck. We are now, it seems, friends for life.

The squally shower threatens to accompany us across the water, but Nigel is more than equal to the task, accelerating smoothly through it and out the other side, and soon we're coasting in bright sunshine, hugging the coastline. The water here is the kind of deep blue you can get lost in. I want to say it's glassy smooth, but it never is. Not really. Even the millpondest surface is interrupted by little ripples and eddies disturbing the surface like fork marks in Christmas-cake icing.

Adjusting my binoculars, and trying to steady myself against the gentle rocking of the boat, I scan the cliffs. I'm familiar with this feeling. It's so easy to think the wildlife will present itself as if on a plate, ready to do our bidding, but its existence doesn't revolve around our entertainment, and against a large and varied backdrop such as this cliff face, even a bird as massive as an eagle can blend into the background. Gulls and gannets, pristine white against the bright blue sky, are easily visible, and shags, their goose-like forms skimming low over the water, catch the eye.

But no eagles.

Sheer cliffs, lone trees clinging to rocky outcrops, all offset by the lush green of the surrounding hills and the chiaroscuro of the cliffs. This is the kind of scenery that holds you entranced, eagle or no eag— FUCKING HELL A BIT OF THE CLIFF JUST TOOK OFF AND IS FLYING TOWARDS US.

A flurry of excitement ripples through the paying punters. But no sooner has it taken off than it's landed again, and a few of the party, looking elsewhere, need a bit of help finding it.

Jamie – young, eager to help – is on it.

'You see the three caves. Find the right-hand one, go right a bit more, then straight up. About a third of the way up there's a lighter patch of rock, then to the right of that there's a tree, and just next to tha—'

'OOH! I'VE GOT IT! THERE IT IS! PERCHED ON THAT ROCK! YOOHOO!'

She's waving, as if to an old friend departing on a long sea voyage, her frantic gestures accompanied by the jangle of jewellery.

I like enthusiasm. My experiences with seasoned birders have sometimes led me to yearn for more overt displays of it. But right now I would kill for the dour and dogged silence of a bird hide.

'Clap your hands! Maybe it'll come down! Hello, eagle! Over here!'

Even at a distance it's easy to see the disdain in the eagle's eye.

A splash off to our left. Another tour boat is throwing fish into the water to lure the eagle towards them. They're thwarted by a pair of herring gulls, opportunists *par excellence*. The eagle remains steadfastly glued to its outcrop.

Any disappointment at the immovability of the bird is soon overcome by the appearance of a second one, apparently from nowhere. And now there are two more of them, soaring close to the cliffs, languid flaps interspersed with glides, white wedge of tail splayed so wide I can count the feathers, illuminated like a nativity scene by the angled sun. Eleven of them, arranged in geometrical perfection.

The bulk of these birds. Thick, barrel chest, planks for wings. Hooked yellow bill, enough to deter any encroachment. Their slow, measured wingbeats disguise a deceptive ability for speed, which we see clearly as one of them circles the boat at a respectful distance. There is a rough wildness to them, in keeping with the craggy rocks that are the backdrop for their display. They are fearsome, majestic, awe-inspiring beasts.

They even manage to silence Ms YooHoo.

They give us fifteen minutes of their time, then they glide back up to the cliff and perch in a way that says the show is over. Nigel steers the boat across the water towards the neighbouring island of Raasay, where we delight in the more sedentary joys of a group of twenty or so seals ('Hello, seals! Cooee!'). They adorn the rocks with a strange kind of lumpen elegance, occupied with nothing more than the onerous duty of being seals. One of them rolls off the rock into the water ('Ooh! 'E plopped right in! Did you see?'), another lifts a flipper in what we take to be a greeting of sorts. Nigel turns the boat round, and we putter back to Portree.

We say our thank yous and disembark. Nigel has been an excellent, informative and generous host. Ms YooHoo remains bubbly to the last, and in the end the purity of her excitement wears me down. The slightly fixed expressions of our fellow passengers when she's in full spate tell me I'm not the only one who thinks she's a bit much. But I also realise that my curmudgeonliness is less than generous, an example of the very dourness of spirit I deplore in others.

She could have spent the day on the sofa watching telly, or in the hotel bar drinking piña colada, or any one of a hundred other things. But she was there. And she loved it. And if you can't show your enthusiasm for something like an eagle, then we might as well give up right now.

If it seems strange to want to escape the hubbub of a relatively deserted place like Skye, then I can only blame it on the camper vans. They are, as previously advertised, relentless in their fecundity. Wherever we go, there's another one, trundling towards us, or lurking behind a boulder like the sinister truck in *Duel*.

So, in the eternal quest for absolute and uninterrupted tranquillity, we take the twenty-five-minute crossing from Sconser to Raasay, experiencing that hint of adventure, the leaving behind of cares that always comes when you set foot on a ferry. This was what visiting Skye itself used to be like, before the building of the bridge, back when we had time – slow, patient travel.

We explore the island by car and bike. It is, to all intents and purposes, a four-road island. There's the one that goes south, the one that goes east, the one that goes in a circle, and the one that goes north. The first two are very short, and the third as long as you care to make it – we take the fourth one, heading northwards. There are views across to Skye to the left, blue sky punctuated by pockets of unthreatening cloud, the sunlight on the water giving it a soft glimmer, and beyond it the contours of Skye's hills, gentle shades of green morphing to grey and black.

We see five humans, a smattering of birds and a flock of sheep. And the deliberately measured pace of our day acts as a relaxant, slowing everything down to a speed quite unfamiliar to a modern city dweller, but no less welcome for all that.

At Brochel, there is a sign: CALUM'S ROAD.

Calum McLeod, a crofter, lighthouse keeper and part-time postman, thought the footpath from Brochel northwards was insufficient for the needs of the inhabitants of that end of the island. Despite years of campaigning and several grant applications, the council disagreed. So he built it himself.

He had a shovel, a pick, and a wheelbarrow. He also had a book called *Road Making and Maintenance: A Practical Treatise for Engineers, Surveyors and Others*. It took him ten years, and remains a testament to the virtues of dedication, single-mindedness and sheer bloody obstinacy.

I try to imagine the strength required to do this. Physical strength, of course, but also strength of will. Unimaginable, to me at least.

We get to the end of the road and walk further, to an abandoned village where twite chatter and a very tattered meadow brown butterfly sits on bracken, and the air is soft and warm and nothing happens and it does it in the most pleasing way imaginable. And then we come back and sit near the ruins of Brochel Castle, looking out across the still water, the light and the soft breeze and the obscene picturesqueness of the landscape combining to do something to us that should be prescribed on the NHS. Not just the landscape, but the slow passage of time, the sloughing away of cares, the lack of urgency you really get only in a place conducive to such things.

Out on the loch there are two birds, tiny specks to the naked eye, floating on the water. They might or might not be black guillemots. I give them a few minutes' scrutiny through the binoculars. Inconclusive.

Across to the left, a rocky outcrop, sand in colour, with seaweed drapery. A rock moves. Not a rock; a seal. And another, and two more, and now I see there are a dozen of them, rock impersonators one and all.

And so the long day wears on.

Sometimes it's enough just to sit in silence in a quiet place with people of like mind, looking out over water, absorbing the contours and shapes and colours and swells and crooks and hollows, saying nothing, but just allowing the world to drape itself over you, bringing you together in a conspiracy of beauty. Cheese sandwich optional.

10

GETTING AWAY FROM IT ALL

In which our author goes to the margins of land and sea, gets up close and personal with amazing creatures, and falls in with a band of excellent people

I am fifty-four years old and I have never held a storm petrel. What the hell have I been doing with my life?

In my defence, opportunities for close contact with this magical and mysterious bird have been few. I have lived my life inland, mostly in cities; the storm petrel, otherwise known as Mother Carey's Chicken, lives almost exclusively at sea. Our trajectories have never coincided. Until now.

Richard has brought the petrel up from the harbour to be ringed, and now Giselle is holding it. It's calm, unblinking in the red light of our head torches, submitting without panic to the brief ordeal of ringing before it's released back into the night. From the harbour below comes the recorded sound of storm petrel song, played to lure them in.

I say 'song'. It's an evocative sound, described by naturalist Charles Oldham as 'like a fairy being sick'. The purring part of it sounds to me like a Geiger counter, and it's capped off with occasional *skrerks* or *cherrkas* or just outright squawks. So maybe 'song' isn't the most intuitive word to use.

Giselle supervises as Will carries out the ringing, patiently crimping the tiny metal band around the bird's leg with specialised pliers, making sure it's in no danger of causing the bird injury or discomfort. When they're both satisfied the job's been done properly, she turns to me.

'Do you want to release it?'

Yes. And no.

I'm a muggle, in bird-ringing terms. This is proper scientific work, carried out by proper scientific people. Giselle and Richard are the wardens, Will a long-term volunteer. Between them they have ringed more birds than I've eaten bowls of Grape Nuts.* Me, I'm a bystander, here for the Skokholm experience, drawn by the remoteness, the isolation, and the birds. My main priority so far has been to get as close to the action as I can without disrupting it. And now I'm being invited to hold a life in my hands. For a moment I feel the incipient horror of getting it wrong, of somehow, in my nervous, eager clumsiness, destroying this tiny, fragile thing.

But I also really, really want to do it.

'Yeah, OK.' All casual like.

* I have eaten many bowls of Grape Nuts.

She gives me the petrel. I cradle it gently in cupped hands, desperate not to squeeze too tight, but equally keen not to let it fly away before I'm told. It fits snugly in my palms, even smaller than I'd expected. It is warm and soft, still, trusting. There's no flapping, no scrabbling to get away. Unexpectedly, it smells – a distinctive, pleasant, musky scent.

If you had to choose a seabird for this first 'bird in the hand' experience, the storm petrel would be as good a choice as any. Gulls are feisty, and the bigger ones do a nice line in beak jabbing; gannets, by turns elegant and thrilling on the water, could have your eye out in a trice; fulmars, larger cousins of storm petrels in the tubenose family, vomit a foul-smelling oily gunk all over you that stays on you for days. I'll take the storm petrel, thanks.

We walk a few yards down the hill, towards the harbour, across the path to the low wall. Giselle tells me exactly what I'm going to do. Place the bird on the wall, keeping my hands around it, then opening them in a V shape, encouraging it to take the righteous path towards the edge and freedom. It might fly straight away, or it might tarry a while. In any case, what we don't want is for it to come back pathside. Not that it would be disastrous; just inconvenient for Giselle and embarrassing for me.

I follow her instructions. The petrel sits on the wall, apparently happy to chill there till breakfast. And then it's gone, plopped off onto the ground just the other side of the wall. Then I hear the tiniest flurry as it flies off into the darkness. I pause for a second, then turn to Giselle, keen to thank her for this lifetime experience.

Before I can say anything, there's a voice from a few yards up the hill.

'Giselle?'

Alice, one of the other long-term volunteers, has something.

'Yes?'

'There's a Manx shearwater on the wall.'

Of course there is.

Giselle turns, her head torch picking out a dark shape. Plump, squat, beaky. It looks up at us, unmoving.

'Correct.'

Welcome to Skokholm.

If you're going to spend a week away from the stresses of life, Skokholm,* a small island off the Pembrokeshire coast, is as good a place as any to do it. Off-grid, and accessible only by weekly boat, it's a mile long, half a mile wide, and home to approximately 200,000 birds. The human population peaks at twenty-six – a mixture of the wardens, volunteers and paying guests like me – so those seeking a hectic social life needn't bother. It's all about the birds – with the odd rabbit, seal or dolphin thrown in.

At one end, a lighthouse; at the other, the cottage, old farmhouse, and associated buildings, all pristine white; bird hides dotted around the island complete the roster of buildings.

Some might regard this kind of isolation as hell – for others, it is paradise. For naturalist Ronald Lockley, who leased Skokholm from 1927 to 1941 and established Britain's first bird observatory here in 1933, it was his dream island.

* I had always imagined it to rhyme with 'Stockholm', and it does indeed have the same root, but it turns out it's more similar to 'Slocombe'.

Lockley had been looking for an island to live on all his adult life. Entranced by nature from an early age, he first tried his hand at poultry farming, but the lure of the wild was great – he and his wife Doris spent their honeymoon on Grassholm,* a neighbouring island inhabited mostly by gannets – and when he discovered that Skokholm was available, he didn't hesitate. The tumbledown nature of the buildings didn't bother him. He and Doris were enterprising and capable, and there was plenty of driftwood about the place to help them patch things up. They were helped in this endeavour by the running aground of the schooner *Alice Williams* in 1928. Lockley paid £5 for its salvage – a bargain price considering how much coal was on board, never mind the rich pickings in the driftwood department. They also salvaged the boat's figurehead, which now takes pride of place in the dining room, keeping watch over all the island's visitors as they eat their evening meals or touch base for a mid-morning coffee before heading out for more birding.

It is, of course, the birding I'm here for, and in particular the Manx shearwaters and storm petrels for which the island is known. Six weeks earlier I'd seen stuffed specimens of both birds on display at the Natural History Museum in Tring. The storm petrel, supported by a stout wire, angled its head towards me; next to it, a Manx shearwater, about twice the size, seemed to be acknowledging my presence with a respectful inclination, as if in the presence of royalty.

But that brief introduction to static, dead birds behind glass was no kind of preparation for the reality of them. The feel of the storm petrel in my hand, the smell of it, the look of its dark eye, its calmness in the face of the unknown, its uncanny wildness –they stay

* Coincidentally the first RSPB reserve in Wales, established in 1948.

with me. If I'm used to seeing birds at various degrees of separation – from the blue tits at the garden feeder to the far-off maybe-black-guillemots on Raasay – close physical contact, however brief, takes my relationship with them to somewhere different. And it's made more memorable by the knowledge that here on the island we are in the crossover zone. These birds come here only to breed; humans barely come here at all. Skokholm is one of the few places where our worlds collide. Their true milieu is on and over the water, where we can never get properly close to them. It makes the sense of privilege, of intimacy, even stronger.

It's an intimacy that was familiar to Lockley, whose work for a long time was the close study of these enigmatic birds. He was exceptional at it, building the big picture by studying the small, and writing about it with a blend of scientific rigour and vivid imagination – including occasional moments of surprising anthropomorphism. His monographs on the storm petrel and Manx shearwater are based on close study of the birds' behaviour over many years, nearly all of it on Skokholm.

The combination of island life and bird migration was one thing – but the presence of a thriving rabbit population on the island also offered an opportunity to make a living from the land. But his idea to breed chinchilla rabbits was short-lived – they proved more difficult to catch than he thought. He also realised that writing offered a more stable source of income. He wrote and he wrote and he wrote. Articles beyond measure, and more than fifty books. As well as birds, he wrote about Skokholm and other islands, but his most famous publication is *The Private Life of the Rabbit*, written in 1964, some years after he left Skokholm and set up a nature reserve at Orielton in southern Pembrokeshire, where he subjected the rabbits to the same keen examination he'd given the seabirds. It's still regarded as one of the definitive works on the species, and was a strong inspiration for his friend Richard Adams's *Watership Down*.

If Lockley's literary legacy is rich, his ornithological influence has been even more potent. His vision of a place dedicated to the close study of birds led to a network of observatories around the country – twenty of them now, from Fair Isle to Portland – where seabird populations are monitored and a wealth of data is gathered to help us understand more about the mysteries of bird migration.

The storm petrel I held in my hand was one tiny part of that, a single pixel in a mammoth picture, built up over ninety years, and still continuously in the making.

It's nice to play a part, however small. It makes you feel involved.

My visit to Skokholm coincides with a comparative lull in the island's year. Its most glamorous breeding visitors, puffins, have gone, the chicks fledged and out at sea, where they spend the majority of their lives. September and October will be migration season – a time when sightings of rarities hit their peak. But the last week of August is relatively quiet. Suits me.

The week before the trip is spent obsessively checking I have all the right gear and enough food. Once you're on, you're not getting off until the next boat comes a week later.

I cautiously tell people where I'm spending the week.

'Oh, it's a small island off the Pembrokeshire coast.'

'Sounds lovely. Will you be spending all your time on the beach?'

Yeah no. Skokholm, it's safe to say, isn't a beach kind of place. There are cliffs galore, but no beaches. Looking at the map sent in the information pack only whets the appetite. Names like Devil's Teeth, Mad Bay, Wreck Cove and Wildgoose Bay give you the impression you're about to embark on a pirate adventure or a *Swallows and Amazons* re-enactment. Quite apart from anything else, it'll be good to be in a place where people don't give you a funny look for wearing a pair of binoculars round your neck.

I'm sharing the island with just a handful of people: there's Pete, all beard and glasses and camera lenses, a ready smile on his face; Chris, silver-haired, bronzed, rugged and capable, the kind of person you could imagine knows how to skin a porpoise and fashion it into a sturdy backpack; the long-term volunteers, Jodie, Will and Alice, whose enthusiasm and youthful energy are stealthily infectious; Mike and Dale, getting on with the monumental task of trawling and collating the Skokholm archive; and Richard and Giselle, who oversee the whole thing and quietly make sure they include everyone. Through them all island energy flows.

Richard gives us the walkaround on the first day.

The Skokholm landscape is picturesque and rugged. One part of the island is given the feel of a soft, bumpy, fantasy landscape by mounds of thrift and moss; clumps of sea campion drape themselves demurely over the entrances to rabbit burrows; round the corner, huge slabs of red Devonian sandstone breach the water at an angle, waves crashing and frothing at their base; everywhere the rocks protruding through the island's surface are speckled with lichens – smooth, mustard-yellow patches rubbing shoulders with hairy green-grey tufts; there are large areas of bracken, turning from lush green to orangey copper, fronds curling in the sunlight; and everywhere there are birds. Fulmars, surfing the air on stiff wings with staggering control; mixed rafts of gulls on the sea;

choughs – small, glossy black corvids with blood-red beaks and legs – hopping about after each other, their twanging calls echoing around the rocks.

Views towards the mainland take in Skokholm's sister island, Skomer, the lurking outline of St Ann's Head on the mainland, and beyond it the oil refinery at Milford Haven, the building of which was vigorously and unsuccessfully opposed by Ronald Lockley. Look the other way and there is nothing but sea, interrupted occasionally by the Rosslare ferry and, today, by seven grey herons flying aimlessly around the lighthouse.

Richard points them out as we follow the path to the quarry at the far end of the island. He's fresh-faced, affable, and possessed of a nice line in understated wit. He tells us about the island's history, its bird population and the life of the wardens. They're on duty from February to November, seven days a week, come hell or high water – and in one of Skokholm's regular storms, there's plenty of both.

The ground is dotted with small slabs, each with a number painted on it. Richard stops by one of them.

'All around this bit of the island is a network of burrows. Homes to rabbits, of course, but also . . .' he moves the slab to one side and reaches into the burrow '. . . to birds.'

He extracts the contents of a Hoover bag and dumps it on the grass.

On closer examination, the grey entity in front of us turns out to be a Manx shearwater chick, but the confusion is understandable. There is a head at one end, tipped with a slender, hooked beak. A wingtip peeks out at an angle at the other end. But otherwise it is just a mass of shaggy grey fluff. You would have to have a heart of stone not to find it overwhelmingly cute.

It sits patiently, plumage more ruffled than its temperament, as Richard gives us a rundown of what life holds for it.

'In a few weeks this straggly grey fluffbundle will develop actual feathers, and the bird will start to emerge from the burrow and get used to the world. At the moment its parents are still feeding it, but at some point they'll head out to sea, abandoning it to fend for itself. And then, one night, it will fly that way' – he points out to sea – 'for a bit, then it will turn left and it will fly to South America, never having made the journey before, and with nobody to show it the way.'

He pauses for a few seconds to allow us to appreciate the impact of this information. And it's while I'm thinking of the instinctive and unquestioning bravery of young shearwaters that a large bird of prey flies quickly over our heads and behind the lighthouse. My view of it is fleeting, but the impression I get is enough for the word 'goshawk' to jump into my head, albeit accompanied by a floating question mark. I daren't say anything, though. I am by some distance the least experienced and knowledgeable birder on the island, and it wouldn't do to lead people up the garden path on the first day.

To my intense gratification, Richard's reaction matches mine.

'Was that a goshawk?'

And at this point he is faced with a problem. If this is indeed a goshawk it would be a moment of some excitement, so his birding instinct is to get after it as quickly as possible. That he would be leaving his newly arrived guests in the lurch is incidental – we can fend for ourselves for a minute or two, or even join him in the pursuit – but there is a flightless ball of raw meat on the grass next to him, and any number of hungry gulls in the area. Leaving the shearwater chick undefended, even for a short time, would be an act of wanton cruelty.

You haven't lived until you've seen an affable Yorkshireman bundle a Manx shearwater chick into a nesting burrow. The chick doesn't even have time to be affronted. And now Richard's away, bounding up the rise towards the lighthouse, and we're after him, not wanting to miss out. We reach the top of the rise just in time to see the bird swoop out of sight again behind the cliff. I have time to see it in more detail, and even though my experience of goshawks is restricted to a single bird in Speyside in 2016, this time I see enough to stifle excitement.

It's a sparrowhawk. A female, notably large, but a sparrowhawk nevertheless. Excellent birds, sparrowhawks, and always a thrilling sight. But not what we hoped it might be.

Richard shrugs.

'Ah well. Not a gos, then. There you go.' He turns to me. 'Didn't it look massive?'

I nod. It did, to me. And it's somehow reassuring that even someone of his knowledge and experience can be fooled as well, even if only for a moment.*

The walk continues, and soon Richard is holding up a dried-out Manx shearwater carcass.

I've noticed them lying about the place, at various stages of decomposition, like dirty socks on the bedroom floor. One was a recent kill, its neck and throat ripped open, exposed to the elements, head impossibly bent underneath the body. This one is long gone, fossil-like in its desiccation.

* I discover later that it would have been just the second record of the bird on the island.

'So this is a dead Manx shearwater. It was most likely killed by the island's apex predator, the great black-backed gull, or GBBG, or "jeeb" for short.' He gestures towards a jeeb standing not far from us – huge, beaky, emitting an aura of barely repressed sullen fury. 'People travel thousands of miles to see apex predators. Lions, tigers. They could just come here and look at a jeeb. They're magnificent.'

Given the low status of gulls in the public's estimation, I suspect he knows it's a pipe dream, but we take his point.

'If you see a dead bird on the island – and you will – just have a quick look to see if its wings have been clipped. We do that to make sure we don't count the same bird twice.' He jiggles the wingtips between thumb and forefinger. 'If they have, that's fine, and get on with your day, but if they're pristine, then just tip us the wink, tell us where it was, and we'll come out and make sure it's recorded. We record all the wildlife we find on the island, dead or alive. At the end of the day you're welcome to join us in the cottage for the log, a roll call of everything seen here. So do feel free to keep records if you want, and share them with us at the log. You don't have to, but all information is good information, no matter how mundane you think it is. In your back garden, a blue tit is an abundant and commonplace bird. On Skokholm it would be a massive deal. And even if these things seem insignificant to you, they're all adding to the big picture.'

He throws the shearwater carcass to the ground, near a small patch of mushrooms, their pertness a mockery of its eternal stillness. And as always when I come across the stark finality of death in nature, the words of Werner Herzog, film-maker and noted Pollyanna, float into my head: '. . . there is some sort of harmony. It is the harmony of collective murder . . .'

It's good to have these reminders of the part death plays in the big picture. We're told over and over again of nature's benefits. And I subscribe to that view, broadly. But so much of it is airbrushed out. And in our keenness to get people connected with it, unrealistic expectations are set. Sometimes the signs at nature reserves, designed to encourage kids to get involved, smack slightly of desperation.

'Hey kids! This is FUN! Is everybody having FUN yet?'

Death isn't fun. It's brutal, impassive. But it's part of it all, and to hide it seems disingenuous. Besides, children love a good kill. Show them a peregrine stooping on a pigeon at 200 mph and their attention is guaranteed. And you can bet that a class of ten-year-olds would look at gulls in a different way if they saw a jeeb, as it is perfectly capable of doing, swallowing a live rabbit whole.

We head back, Richard continues with the talk, and as we make our way round Lockley's Dream Island, its magic revealing itself to us in bright, breezy sunshine, I fall ever so slightly in love with the place.

Night falls. The isle is full of noises. It would be a stretch to describe them as 'sweet airs', but they do give delight.

They are the sounds of Manx shearwaters coming in from the sea – hoarse, rhythmic wailings, like haunted asthmatic wood pigeons.

They spend the day out at sea, feeding, coming in to land only when their predators are safely tucked up in bed. Evolution has decided they must be strong flyers, adapted to the low skimming over the ocean swell that gives them their name. So they have slim wings and powerful bodies. But this aerial strength has a side effect:

they are comically ungainly on land. Their legs are placed far back on their bodies, so to move on land they either do short stuttering bursts on tiptoe, or shuffle along on their stomachs, pushing with their wings and feet. The young ones, exploring the strange world outside their burrows for the first time, can barely do that. This makes catching and ringing them relatively easy, so after roll call, a party heads out along the main island path that runs from the cottage to the lighthouse and gets to work.

There are Manxies everywhere. Several pass overhead as I follow the ringers. One blunders into my legs and scuttles off into the bracken. Another sits in the middle of the path, almost defiant in its stillness. Up ahead, the ringing group stand in a huddle, ringing a bird with care and diligence.

Richard walks towards me, a Manx shearwater tucked under each arm.

'Hi, Lev.' He inclines his head towards a bird shuffling slowly along the edge of the path. 'Grab that Manxie, will you?'

Yes. Yes, I can definitely do that.

It turns out I can.

The trick to picking up a Manx shearwater, I discover, is to move slowly, then fast. Approach with care, making sure you don't spook it. Then make your move, ensuring you hold its wings firmly to its side.

I share this knowledge just in case you might at some stage need it. You never know.

I bend down, grab it, and clutch it to my bosom.

Is it stressed by this capture? It's impossible for me to tell. It shows no resistance at all. Its only response is when I relax the grip of my left hand and it slips into a slightly less comfortable position. A wriggle or two, an irritated peck, a readjustment, and all is well again.

There's a kerfuffle off to the right. A Manxie, emerging from the bracken as fast as its wings can carry it. It flies low and fast, straight towards Chris. Half a dozen head torches turn towards him.

He barely blinks, reaching out and catching the bird in both hands like a rugby ball, then holding on to it in the approved manner, nonchalant as anything, as if he doesn't consider a day complete until he's caught a fugitive seabird.

He hands it over to Richard.

'There you go. One Manxie.'

Signed, sealed and delivered.

By any objective measure, sea watching is a ridiculous thing to do – hours are spent staring into the roiling expanse, trying to distinguish one small black dot on the surface of the water from another. And yet there's a mesmerising magic to it as well, a zen-like mixture of concentration and meditation as the sun and foam and waves and swell and spume conspire to send you into a state of extreme serenity.

A nagging wind blows in from the north west, just strong enough to be called 'fresh'. There is some high cloud over the mainland to my left, but otherwise the skies are blue. Sunset is a couple of hours away. I'm trying to work out if the raft of auks in the distance are guillemots or razorbills. Matters are made more difficult by the swell, which obscures them from view just as I've fixed on what I think is a salient feature. With the naked eye, they're invisible; through the telescope they're black and white blobs bobbing on the surface, the specific arrangement of the black and white being a key clue to their identification. My confused brain ties itself in knots.

I leave it, sweep the horizon with my binoculars, and work my way back across, picking up a few gannets in the middle distance. And now, as I accustom myself to what at first seems an empty, featureless seascape, I see something else. Narrow parentheses low over the water, describing long arcs this way then back and this way then back and this way then back, the swaying rhythm of their glides almost lulling me into a trance.

You wouldn't believe they're the same bird. Last night, stranded on the path, shuffling out of their burrows, ungainliness in bird form; today, over the sea, masters of their elements, graceful, easy, flowing. I try to equate these slender creatures with the dumpy thing I held in my arms eighteen hours earlier. It's a stretch.

There's a rustling without. A cheery bronzed face appears at the window.

'Oh. Hello, Chris.'

'Hello. Anything about?'

It feels rude to the seabirds I've been admiring for the last half-hour to say, 'Nothing much', but convention dictates that we don't get excited about the usual stuff.

'Oh, just a few gannets feeding out there, a couple of fulmar, Manxies beginning to come in . . . and there's a raft of guill—'

'Oh look, there's a sunfish.'

He's pointing down at the sea, close in, just over the lip of the cliff. A few juvenile gulls are floating gently on the surface. I'd seen them and then ignored them. Identification of juvenile gulls is best left to the professionals, as far as I'm concerned, and I had enough on my plate what with the guillemots/razorbills, not to mention the simple and often underrated act of just switching off and letting the mind wander.

Now I look in the direction he's pointing, and I see, floating just beneath the surface, an indistinct white shape. Closer inspection reveals a fin or two. It wasn't there when I last looked, a few seconds earlier. But even if I had seen it, I'm pretty sure I would have thought it was a plastic bag. It's drifting gently across our sightline, closely accompanied by one of the gulls, which gives it an exploratory peck every few seconds.

'I caught one of those once on my surfboard. Just floating along and it jumped up into my arms.'

Of course it did. Chris has form in the wildlife-catching arena.

The sunfish, still beset by gulls, rotates slightly in the water and hoists a fin languidly above the surface.

'Oh hi, Chris,' it seems to be saying. 'How are things? Who's your idiot friend?'

I'm aware that this feeling of irritation stems only from my own sense of inadequacy in the face of Chris's extreme competence. He's not trying to be more outdoorsy than me, or more attuned to nature. He just is. And if I have difficulty with people who can catch flying Manx shearwaters as easy as winking, then that's my problem, not his.

It doesn't help that he's really nice.

The sunfish has drifted out of view, and the gulls look slightly peeved at the disappearance of their new toy. I redirect my attention to the Manxies and their eternal sweeping dance.

A new morning. Fresh and bright and dewy.

Four days in, and I've developed a routine. Up early, straight down to the harbour to say hello to the seals.

There are eighteen of them today, lolling about on the rocks or bobbing languidly in the water, whiskery noses poking up above the surface, their eyes deep dark holes which only encourage Herzogian thoughts. The natural melancholy induced by the sight of them is enhanced by the acoustics of the harbour, which amplify their sounds – eerie, otherworldly howlings – in an uncanny way.

For all that they effectively do nothing for half an hour, this is time well spent, and I allow myself to stay still and savour it for a few minutes more. My instinct is to keep moving, but one phrase from a conversation with Richard the day before sticks in my head.

'Sometimes all you need to do is sit back and let it all come to you.'

It's not meant as a rebuke, but it strikes a chord. Activity is all very well, but I sometimes – often, in fact – forget the benefits of stillness.

On the other hand, I'm hungry. I head back to the observatory. The kitchen and courtyard are empty. Everyone else is out and about. I make myself a coffee and sit on the bench, looking back towards the mainland – no more than three miles away, but effectively on another planet. Time, ever variable in its passing, slows to a standstill. Peace washes over me. A painted lady butterfly flutters past, lands on the buddleia next to me. Every few years there is a mass influx of these large yet somehow delicate butterflies from the Continent. This has been one of those years, and I have drunk my fill of them while on Skokholm. Part of their attraction is the light orange and black colouring – less gaudy than the more obvious charms of a red admiral or small tortoiseshell – but it's the migration that bewilders. My brain can, more or less, encompass the idea of a bird undertaking those long, hazardous journeys, but a butterfly? Show me a hundred times and I still won't quite believe it.

Mike leaves the library and walks past me towards the nets on the left. This is a Heligoland trap, named after the place where they were invented. Built like a fruit cage, they have a wide opening at one end, narrowing towards a catching box at the other. Once in the cage, birds funnel towards the box, where they are easily retrieved for ringing.

Mike is one of those people whose eyes tell the story. A favourite uncle or grandfather, quiet humour just below the surface, intelligence and wisdom ingrained. He says little, but understands much.

He lopes back past me, into the courtyard, towards the ringing hut. He has a cotton bag in his left hand. It is wriggling.

'Would you like to see me ring a whitethroat?'

I think so, yes.

He handles the bird with an expertise born of decades of experience, talking me through the process. Log the bird. If it is already ringed, take the ring number; if not, apply the ring using the appropriate ring and pliers. If fitted correctly, the ring is no inconvenience to the bird. But care must be taken. The number one rule for any ringer is that the bird's welfare comes first.

There are 2,300 trained ringers in Britain and Ireland. It's a rigorous process, involving many hours of supervision by a ringing trainer, and many early mornings, this being the most active time for bird movements. It takes several years to qualify, and even then you are constantly monitored, and have to remain active as a ringer to keep your licence. You can become prime minister with fewer qualifications.

The whitethroat looks tiny in Mike's hand. He applies the ring with efficiency and minimal adjustment, logs its details, and the time and date of the bird's capture. He measures the bird's body length and wing feathers. Then he pops it into a little tube and weighs it. He blows through a short straw, ruffling the bird's chest and stomach feathers so the dark pink flesh underneath is visible. His movements are small, neat and controlled.

'I'm measuring how much fat the bird has. Too little, and it'll likely not survive migration; too much, and it might slow it down and make it easier prey for predators on the way.'

Soon enough, he's done, and now he looks at me, the twinkle ever present.

'Have you ever held a small bird in the hand?'

He knows the answer already.

'Just the storm petrel the other night.'

He describes how I'll need to hold it, using the 'ringer's grip'. The bird will be held loosely but firmly, its neck between my index and middle fingers, the rest of it in the palm of my hand.

He gives it to me. The whitethroat is calm. Whether this is because it is resigned to its fate or because I ooze trustworthiness, it's hard to tell, but I opt to believe the latter. It looks at me with a cool black eye, completely still. I would like to keep it forever, or at least a minute more. But it is a free, wild thing, its enslavement temporary.

I bring my left hand up to my right, forming a little cup. The whitethroat's feet touch down briefly in the palm of my left hand. A tiny scrape. And then I lift my right hand and open up the fingers, and it's free to go. For the merest instant, so brief I can only conclude it's wishful thinking on my part, the whitethroat stays in my hand, as if it too would like to prolong our relationship. And then it flies, wings strong, fast and delicate on the air, and the feeling, so vivid, is already a memory as it makes for the bush on the other side of the courtyard, the only rebuke to my tenderness a bathetic squirt of poo on the cobbles below.

Life-changing moments take many forms.

The last night. Roll call.

Richard goes through the list in taxonomic order as usual, inviting contributions from the floor. Over the course of the week I've become confident enough in my knowledge of the island's geography and the likely make-up of its bird life to be able to contribute some definitive counts. Gannets, fifteen at Howard's End; buzzard, one at lighthouse; rock pipit, thirteen at Isthmian Heath; swallow, ten over North Plain.

I leave the gulls to the experts.

Occasionally a mild discussion breaks out, delaying proceedings. Richard sits it out, takes a sip of whisky, allows the chatter to die down, then nudges the room back to the business at hand.

'Guillemots, anyone?'

Forty-eight on the sea at Crab Bay, apparently.

There is a quiet warmth to these proceedings, a community spirit engendered by the wardens and springing from the reason everyone is here – a love of nature in general and birds in particular. Skokholm visitors come in different shapes and sizes – birders, artists, general wildlife people, photographers, people just wanting a break and not interested in the run-of-the-mill, and many more. And they come back time and again. This is in part down to the lure of the place. Of course it is. But there's more to it than that.

The trick that Richard and Giselle have pulled off here is a remarkable one. Everyone is welcome. It's as simple as that. If you come to the island, you belong. And you're not just welcomed; you're positively included. It's a rare and special thing, and as Richard winds up proceedings, I feel a pang, knowing this is the last time. This is a place where worlds meet. Land and sea, day and night, birds and humans. I'm leaving it sooner than I'd like.

Perhaps I'll return. Yes, that's a plan.

Outside, it's clear and chilly with a hint of melancholy. We head out along the lighthouse path. The relentless cycle of ringing continues up ahead, but I trail behind, wanting a bit of solitude before lending the mighty weight of my expertise to the ringing effort. A Manx shearwater looks up at me from the edge of the path, almost expectant. I pay it no heed and stroll on. I find a rock, as flat as I can, and lie on it. One of Skokholm's many charms is the darkness of the night sky. I look up at it, not looking at anything in particular. There is no moon, and the stars are all the brighter for it. There are a lot of them too – far more than I've ever seen in London. I don't know what they all are, of course. I can do the Plough, and Orion. And I suppose if I had to, I could find Polaris. I might know the names of other celestial bodies – Cassiopeia, Cygnus, Canis Major, and maybe a few others not beginning with 'C' – but I'd have zero chance of finding them. My knowledge of the night sky is vague, to say the least.

And as I look up at them and realise the full extent of my ignorance, something goes click in my head.

I have one more place to visit. Maybe two.

11

TO INFINITY AND BEYOND

In which our author gets an inkling of what it's all about

'Nature' is a human construct. And its definition varies depending on who you talk to.

Ask a random passer-by, 'What is nature?', and they will most likely talk about the countryside and trees and birds and *Springwatch* and did you see that programme about the iguana and the snakes?

Ask a botanist and they will talk about flowers and stamens and chlorophyll and probably photosynthesis, because that is after all the very stuff of life.

Ask a particle physicist and soon your head will be full of cotton wool because they tend to bring things like muons and gluons into it, and insist that if you think you understand it you don't in fact understand it, and the very fact of even thinking about it means that you've changed it,* and it all goes around in your brain until you're a great big sobbing mess.

Say it to an astronomer and they will direct your gaze upwards, not to the great spotted woodpecker drumming at the trunk of a tree, nor even to the cumulonimbus clouds laden with water to dump on your head, but up and beyond to the moon and Ursa Major and Betelgeuse and Orion, and there will be numbers, extremely large ones, and even though you nod and say, 'Yes, I see', you don't see, not really, because it is too big for the human mind to truly comprehend.

Different people see different things.

On 16 July 1969, three adult male humans inserted themselves into a metal box and were propelled into space. They did it, I have to keep reminding myself, voluntarily, and they did it in the interest of human exploration.†

That eight-day journey – 238,900 miles, give or take,‡ and the same, more or less, back again – was at the time the longest journey away from earth ever undertaken by humans.§

* I might not be representing it entirely accurately, which in a way proves my point.

† Yes yes, I know – it was also in the interest of a rash promise their president had made a few years earlier, and in the interest of beating the commies to it, and in the interest of a technological race that was acting as a proxy for nuclear war, but *apart* from that . . .

‡ The distance varies between 225,623 and 252,088 miles, depending on where the moon is in its orbit.

§ It was surpassed by Apollo XIII, which swung round to the dark side of the moon the following year.

Fifty years later, to the day, I set up a telescope and point it towards the night sky. It's the spotting scope I use for birdwatching – good for things a quarter of a mile distant, less good when it's a quarter of a million – but it's better than nothing, and will have to do.

It does.

I don't need an excuse to look at the moon. It's the second most visible object in the sky, after all, and there are times when I find it difficult to look away from it, such is the hold it exerts. But tonight I have particular motivation to pay attention to our only natural satellite, our child, formed (we think) 4.5 billion* years ago from the debris of a collision between earth and a Mars-sized planet called Theia.

In one of the cosmic coincidences we love so much, the fiftieth anniversary of the Apollo XI launch also saw a partial lunar eclipse. It really would be rude not to, not least because it gives me an excuse to use the word 'syzygy' – any word that is two-thirds composed of the final two letters of the alphabet is fine by me.

An eclipse happens when the sun, earth and moon are in syzygy.† With solar eclipses the order is sun–moon–earth; with lunar eclipses it's sun–earth–moon. Solar eclipses are more dramatic and glamorous, their particular allure being the result of another cosmic coincidence, the kind that might just persuade me of the possibility of the existence of a supreme being. It is this: the sun is approximately 400 times further away from earth than the moon, and the moon's diameter is approximately 400 times smaller than the sun's.

* A word much improved if you say it in the voice of the great Carl Sagan.

† Having used it, I should probably explain what it means. It means 'configured in a straight line' when applied to three or more celestial bodies.

So the two objects look the same size in the sky – if the moon were bigger or smaller or closer or further away, this wouldn't be the case – and when one is in front of the other the effect is, well, dramatic and glamorous. As we have just the one moon,* this is a coincidence worth recognising and celebrating.

The advantage of lunar eclipses over the solar ones is that they happen during the night and you can observe them directly without blinding yourself, and of course in Britain they are far more common. Even then, they're infrequent enough to make even the part-time skywatcher murmur, 'Ooh, must remember.' Total lunar eclipses average out at about one a year; throw in the partials and penumbrals and the count nearly triples.

I happen, on 16 July 2019, to be on the Isle of Wight, in an area of comparatively low light pollution.† That the moon rises over a calm sea only adds to the occasion. And the complete lack of clouds impeding my viewing can be attributed only to the power of prayer. I've been begging for a cloudless night for a week.

I track the moon's progress across the clear night sky, examining its craters, valleys and mountains as closely as possible. I've done enough homework to know that the clearly visible crater, low down on the moon's surface when viewed from the northern hemisphere, is named Tycho, after the Danish astronomer Tycho Brahe,‡ and

* Unlike Jupiter, with its frankly ostentatious total, at last count, of seventy-nine.

† Compared with London, that is, which is admittedly a low bar.

‡ I haven't collated all the information, but he belongs to a strange subset of humans who have been exhumed twice. In his case the two exhumations, in 1901 and 2010, were to establish the circumstances of his death (most likely a burst bladder), but also to establish the material from which his artificial nose was made (brass, not gold as was believed during his lifetime).

I focus for a bit on its silvery luminescence, slightly dulled by the encroaching shadow – our shadow – moving across the top.

I follow one of the tracks leading from Tycho, up and to the right, and my eye lands on a dark area. I zoom in on it.

The Sea of Tranquillity, recipient of Neil Armstrong's small step half a century ago.

When you know, it's the first thing you look for. Otherwise it would just be another dark patch, the wonky nose between the Man in the Moon's eyes. But I find myself seeking it out, and it's almost impossible to look at it without hearing the crackling audio recordings, picturing the tiny capsule, imagining the thoughts and feelings and experiences of those three men. I try to grasp the feeling, the simultaneous smallness and hugeness of that act, but it's slippery, as so often when I try to contemplate things beyond the human scale. I try to understand it by thinking of the journey in terms of other journeys more familiar to me. From London to Bristol and back is about 240 miles. Do that a thousand times.* Walk around a bit, then do it again.

Or I could simply ask an Arctic tern to explain it to me.

It feels important to appreciate these distances, to try and understand them. For thousands of years the moon was impossibly remote, an awe-inspiring and constant presence in our skies. Then, overnight, the impossible distance became possible. The moon still holds mysteries, but now we look at it and think, 'Yes, we can go there.'

* Obviously the moon rocket went a lot faster and didn't stop for a Gregg's at Leigh Delamere, but thinking about it like this makes it more manageable somehow.

And now Mars is next, the huge distance separating us talked about as a feasible journey, as if, having safely made it to Bristol, we're now planning to drive to Sydney.

I can get my head round these figures, at least enough to be relatively comfortable with the knowledge of our own insignificance, the tininess of our existence in the grand scheme of things. It's freeing, this knowledge, if you allow it to be, releasing the psyche from existential angst and diverting it towards more mundane matters, such as who invented fig rolls, and what did they have against figs?

But soon one's understanding of the scale of things spirals out of control, the sheer insignificance of what we regard as mammoth distances made plain as we contemplate what lies beyond.

Best to stick with what we can manage for the moment.

There's one more feature of the moon's visible surface I've memorised, something I'm keen to locate, or get as close to locating as I can, before the shadow reaches it. I'm not expecting it to be visible through my telescope, but close is better than not at all.

I take a line from Tycho up to another bright spot: Copernicus crater, an island of shine in dark surrounding seas. And then up a bit more, into the broad dull patch called Mare Imbrium,* the Sea of Showers. It's larger, vaguer than the Sea of Tranquillity – about 700 miles wide, the distance from London to Pisa, Galileo's birthplace. In that context, an eight-mile crater is tiny. Maybe it's that little speck just visible in my telescope. Yes. Let's say it is.

* If it seems as if I'm an expert in lunar geography, rest assured I'm not. These are just the ones I know about and have looked up. The rest of it is a mystery awaiting exploration.

It was tempting, when thinking about the great figures of astronomy, to choose William Herschel, the German-born musician and astronomer who was the first president of the Royal Astronomical Society, and, among other things, discovered Uranus.

And then I read about his sister Caroline and regretfully gave William the old heave-ho.

On the spectrum of women whose potential was thwarted by their gender, Caroline Herschel isn't at the most disastrous end. She did at least receive recognition for her work in her own lifetime, notching up several firsts for women in scientific fields, and particularly astronomy. But many of her achievements were attained in her brother's shadow, and while it was thanks to him that she showed any interest in the subject, it's hard not to wonder what she might have achieved if allowed to do more than act as his amanuensis.

When she joined William in England on the death of their father in 1767, most of her time was spent as his housekeeper and assistant. As well as running the household, her duties included helping him in his astronomical activities, doing 'nothing . . . but what a well-trained puppy dog would have done'. She pandered to his every need, even being required to put food in his mouth as he worked, and spending hours polishing the lenses of his homemade telescopes.

As his interest in the subject grew, so did her involvement, but any desire she might have had to pursue her own researches was subservient to his needs. William was the astronomer; Caroline his assistant. It was Caroline who sat at the bottom of the telescope recording William's observations, a task that required speed and meticulous precision, as well as an advanced understanding of astronomical recording methods; it was Caroline who, while William did terribly important work on an astronomical catalogue and double stars and grown-up things like that, was given a smaller telescope and told to sweep the skies looking for comets; and it was Caroline who undertook, at William's request, the monumental task of reappraising, cross-indexing, updating and correcting John Flamsteed's *Historia Cœlestis Britannica,* the pre-eminent star catalogue of the time.

In 1782 William was appointed 'The King's Astronomer', and they moved, by royal request, to Datchet, where he could be on hand to show the king and his guests the wonders of the night sky. While Caroline's role in this new venture was still secondary to William's, she made her mark by discovering eight comets and fourteen nebulae, and in 1787 her efforts were rewarded when she became the first woman to be given an official government position (even if it was as assistant to her brother) and the first to receive a salary for work in astronomy.

Caroline lived to the age of ninety-seven, outliving William by twenty-six years, and moving back to Hanover after his death. Her astronomical efforts there were hampered by architecture – the tall buildings where she lived restricted her view of the night sky – but she spent most of her time helping her nephew John* with a catalogue of nebulae.

In 1828 she was awarded the Royal Astronomical Society's Gold Medal. The next woman to achieve this was Vera Rubin in 1996. She was, by any measure, a remarkable and important scientist. Finding the crater on the moon named after her feels like the least I can do.

* Who, you'll remember, invented cyanotypes all the way back on page 94.

The shimmering reflection on the sea, in recent nights so pellucid, now has a subdued dullness to it, as if sapped of energy by the overbearing shadow of its mother planet. The shadow makes its slow way across the moon, the surface taking on an orangey shade, and soon the lunar-impact crater C. Herschel is swallowed by darkness.

If I think about it I can be thrilled by the knowledge that the shadow inching its way imperceptibly across the moon's cratered surface is earth. That's us, that is. Hello, Mum! But that requires a cognitive leap, albeit a small one. My reaction to the moon's colour is visceral.

Knowing that the reddening of the moon during a partial eclipse is caused by the scattering of light through the earth's atmosphere* is one thing; seeing it happen in close-up – the quarter of a million miles between us contracted by optical magic – is quite another. Knowledge can remove mystery – 'oh, there's quite a simple explanation for that' – but it's possible to be two things at the same time: on the one hand a modern human with a basic grasp of science and on the other a primitive staring in awe at a supernatural and inexplicable phenomenon.

The science itself is amazing; its effect doubles the wonder.

I'm aware, as I look, that I don't have the right tools for the task. My scope, with its 20×–60× magnification, is great for differentiating a gadwall from a garganey at a hundred paces, but if I want a view of the wonders of the Andromeda galaxy, I'm going to need something a bit more high powered.

*

* About 74 per cent nitrogen, 21 per cent oxygen, 1 per cent argon and then a whole load of other stuff including variable amounts of water vapour.

There are about twenty people in the classroom. A mixture of ages. On my left, two children – about eight and ten – sit with their mother, quiet with suppressed anticipation. The excitement of the group to my right, six men and women of my age or thereabouts, manifests itself as hearty banter, slightly louder than strictly necessary. A low buzz fills the wood-panelled room, expectations slightly undercut by the unspoken truth: we probably won't get to see any stars.

Any pre-planned astronomy excursion is hostage to the prevailing weather conditions, and for a week I've been checking the local forecasts, using the time-honoured technique of working through all the weather sites until I find a forecast I like.

There are no forecasts I like. The conditions for the night of my visit to Kielder observatory are resolutely set to 'thick cloud with an extra slice of cloud and a side order of cloud'.

They're prepared for this at Kielder – they have to be – and the enthusiastic and knowledgeable team will deliver an evening of entertainment, whatever the conditions.

I will have a good time. I will learn loads of stuff that will no sooner enter my head than leave it again.

I will almost certainly not do any stargazing.

Kielder observatory sits high on a fell near the Scottish border. It's in the International Dark Sky Park formed by Northumberland National Park and Kielder Forest & Water Park, one of the darkest places in Europe. The nearest A road is fifteen miles away. It is as remote as anywhere in the UK, and on a cold December night it feels like it. The approach road to the observatory, winding upwards through the forest, feels endless, lonely, and just a bit spooky. It's the kind of road you drive up hoping you won't break down, while your mind rehearses all the possible scenarios that might unfold if you did, most of them involving dangerous criminals of one kind or another.

As I get out of the car a chill wind ruffles my hair. By the observatory – a modern timber oblong with twin turrets to house the telescopes – a small wind turbine rotates frantically, emitting a soft, relentless, rhythmic wailing that only adds to the desolate feeling of the place. The constant light and bustle of south London feels a universe away.

The observatory's primary remit is to put on events such as this one, explaining the mysteries of the universe to the general public and giving them a chance to use the observatory's powerful sixteen-inch telescopes to look at the infinite wonder of the night sky. The staff giving the talks are young, enthusiastic and communicative. Good humour runs through them like writing through a stick of rock.

'Just a word about our composting toilet before we begin. Think of it as like a black hole – once something crosses that event horizon it's not coming back.'

The plan for the evening is laid out. As expected, we're in for an indoors kind of time, but we're assured that at the first sign of clear skies we'll be out of the classroom and looking through telescopes quicker than you can say 'coronal mass ejection'. Hayden – genial and softly spoken, with a nice line in dry wit – tells us about the solar system. There are facts and numbers and sciencey titbits galore, pitched at just the right level to make you feel clever for knowing some of it already.

'When we look at the sky, we're looking back in time – just over a second in the case of the moon, eight minutes when it comes to the sun, and years when it comes to the other stars.' He lets it sink in. 'And the more you think about that, the more it messes with your head.'

These are statistics I more or less know, but have never quite come to terms with, so it's wonderfully reassuring to hear this level of awe from someone who spends their whole time considering such things.

He talks about the sun and Mercury and Venus, and then works his way outwards, covering each of the planets and their individual quirks and characteristics in turn. Next to me the two children are agog, not moving a muscle. I glance across after twenty minutes or so and meet the mother's eye. We exchange smiles.

And now Hayden broadens the scope, messes with our heads just a bit more.

'Look up at the sky, and you'll see stars. Our sun is just a star, one of 400 billion in the superstructure called the Milky Way galaxy. We think of it differently because it's close to us, but all those stars have planets orbiting round them, just like ours. And beyond that, there are billions more galaxies.' A short gap, and then he speaks more softly, so softly I nearly miss it, except you can't miss it because there's one of those silences you get only when a roomful of people is listening very closely.

'And here's the wonderful thing. We think of all that as "out there", but of course we're a part of it, and we're all – you and me and everybody – made of the same stuff that makes up the universe. We're all stardust.'

It's the kind of thing, said by the wrong person in the wrong tone of voice, that might have me rolling my eyes. But this is a scientist, not expressing some woolly idea in poetic terms, but baldly stating a scientific fact with great enthusiasm and persuasiveness.

This isn't 'ooh don't you sometimes think the stars are God's daisy chain?' stuff; this is pragmatic and down to earth, yet it contains a profound truth.

We are all stardust.

It's a thought both terrifying and comforting at the same time, so it's probably just as well that we're interrupted before it can take hold.

One of Hayden's colleagues pops his head round the corner.

'It's clearing up.'

A little ripple runs round the room. For all that the talk has been entertaining, we've come here for the real thing. They split us into two groups and we file outside. I find myself on the fringes of Hearty Banter group as we go up the ramp and gather round the telescope. Set in eight metres of concrete for stability, it's impressively large, as you'd hope, and has another, smaller scope bolted to its side. It's controlled by a computer on the far side of the room. Such is the sophistication of the software, you just type in the name of the thing you want to look at, and the telescope does the rest, following the celestial object like Philip Marlowe on the trail of a hoodlum.

M – O – O – N.

The telescope moves smoothly round, pointing through the aperture in the roof that's been opened just enough for the purpose. Naz, our guide for this part of the evening, has a little look through the eyepiece, then moves aside. We shuffle round to take our turn, everyone wanting to look and look and look, but too polite to hog it. There's a strange mixture of reactions. One woman, channelling Ms YooHoo from the Skye eagle trip, gives an excited squeal.

'Ooh, it's ever so bright!'

This triggers a bout of excited joshing in her group. It dies down after a few seconds, to be replaced by a contemplative silence interrupted only by the shuffling of shoes on the floor and a quiet murmur.

It's my turn. As I put my eye to the eyepiece, Naz tempers our excitement with a different perspective.

'The moon's a beautiful thing, of course, but the fuller it gets, the worse it is for looking at the rest of the sky. It reflects only about 4 per cent of the sun's light, but that's enough to mask quite a lot of the celestial objects around it. Some astronomers call it "the devil's lightbulb".'

Just at this moment, I don't care about the rest of the sky. The moon is giving me plenty to look at. It was entrancing enough seen through my spotting scope; now, viewed through this comparative behemoth, it's apparently close enough to touch, and utterly mesmerising, the effect of the silvery glow magnified, each pockmark vivid and real. I look at the craters on the narrow margin where dark meets light, the casting of the shadows rendering them truly three-dimensional. Then I yield unwillingly to the next in line and stand aside, looking up at the clearing night sky and thinking Big Thoughts.

We adjourn for hot drinks and milling. I find myself next to a middle-aged man – ruddy of cheek, genial of tone – who offers his thoughts. They tally almost exactly with mine.

'Makes you think, dunnit?'

It's a simple, clichéd truth. It really does make you think.

We move outside again in small groups. Some go back in to the large telescopes, others opt for an examination of meteorite samples. I head for a viewing platform outside the observatory, where the staff have installed two portable telescopes, easy to manoeuvre so we can look at any part of the sky that takes our fancy.

In a triumph for the discipline of mind over matter, if not for the art of weather forecasting, the clouds have dispersed enough for us to see large swathes of sky. For a city dweller, used to obscene levels of light pollution, the darkness itself is a kind of miracle. Already, as my eyes adjust, I can see many more stars than on even the clearest London night.

Adam, who to my aged eyes looks as if he's way past his bedtime, places himself at our disposal, asking us what we'd like to see. There are shrugs and doubtful looks, possibly because nobody wants to say the obvious. In the end I fling caution to one side. Who cares if I seem gauche?

'Orion?'

He swings the telescope round and unerringly focuses on the three stars of Orion's Belt, moving smoothly into a quick rundown of the most clearly visible stars that make up possibly the best-known constellation in the night sky. With the usual awe of the inexpert for the expert, I can't believe his depth of knowledge, the speed and ease of the answers to any questions we throw at him, and the facility with which he handles the telescope, swinging it round to show us another constellation with speed and dexterity.

But mostly I can't believe he's wearing shorts.

It is not warm. I'm wearing a hat, thick gloves, long johns, woollen socks and four layers under my padded coat; he's swanning around in a light fleece and track shorts as if he's just off for a stroll down to the beach.

'So, if you go up to the left from the belt at roughly ninety degrees, you can see Betelgeuse – Orion's left shoulder. It's often pronounced "bettle-guhze", but then Brian Cox pronounced it "beetle juice" on the telly, so, you know . . .' He shrugs. 'It's gone a bit dim recently, so some people think it might be about to go supernova, but bear in mind that it's anywhere between 430 and 650 light years away, so it might already have happened and we'll only just be seeing it from here. In any case, it'll happen soon, by which of course I mean any time in the next hundred thousand years.' He gives a quick smile, full of humour. His energy is infectious.

He angles the telescope low, as if about to fire a projectile across the ground into Kielder Water a few miles away.

'Have a look at that.'

I have a look. Low in the sky, a star bright enough to be clearly visible with the naked eye. It twinkles like the eyes of a favourite uncle. In the telescope it looks like a miniature glitter ball, red and blue and green sparkles turning slowly around each other.

It looks alive.

'That's Sirius, the Dog Star. Famous star, very bright, the brightest star in the sky. It's a binary star, which means it's actually two stars, orbiting each other. But they're close enough to each other that from here they look like one.'

'How close is close?'

'Depends. Between eight and thirty astronomical units, depending on where in the orbit cycle they are.'

They told us what an astronomical unit is. I'm expected to know. A short pause as I will the knowledge back into my head. Ah yes. The distance between us and the sun. Thirty astronomical units – thirty times ninety-three million miles – seems far, but they're eight light years away, which is like a gazillion miles or something. Everything is relative.

'It's very low in the sky, so the twinkling you see is the scattering of the star's light by the atmosphere, not the star itself.' He turns to me as I step back from the telescope. 'They call it the disco star.'

'Really?'

'Yes, really. It's in all the astronomical textbooks.'

'Oh. So not really then.'

'No. But it would be good, wouldn't it? Now, let's have a look at the Plough.'

Swing, drop, focus.

We call it the Plough, Americans call it the Big Dipper, but I call it the Saucepan, because nobody knows what a plough looks like any more, but they do know what a saucepan looks like, and that, to me, is what it most resembles. When our ancestors named the constellations, they used things that were familiar to them, and the names varied across the world according to local experiences, beliefs and mythology. The night sky was an integral part of people's lives. Not any more. We know lots of stuff they didn't, of course, but that ordinary contact with the world around us has been eroded. In learning, we forget.

At my request, Adam takes us through the names of the Saucepan's stars.

'OK, so, from the end of the handle: Alkaid, Mizar, Alioth, Megrez, Phecda, Merak, Dubhe.'

I try to construct a mnemonic to help me remember. AMAMPMD, the initial letters of each star, doesn't exactly trip off the tongue. Using the first two letters is little better: Almialmephmedu. I decide to remain in a fog of ignorance for the moment.

'How far away are they? I mean, they all look similar brightnesses. Does that mean they're approximately the same distance?'

'Well, sort of. They're all between about 80 and 125 light years away. Dubhe is the furthest. Alkaid, at the tip of the handle, is about a hundred light years.'

'So the light we're seeing now left it a hundred years ago.'

'It's a bit more complicated than that, but basically, yes.'

I look up at Alkaid, its steady one-hundred-year-old light constant against the inky depths, and I can't help the thought that pops into my head, that this light I'm seeing now was created, more or less, at the same time as my father was born.

The complexities of the space–time continuum mean this isn't the whole truth – I'm dimly aware of that, but I failed physics 'O' level, so attempts to clarify it for me are doomed. Besides, I prefer the poetry of the thought to the wonder-diluting reality.

Adam, a human dynamo with cold-resistant legs, has swung the telescope round and is focusing on a different part of the sky.

'Ah! There we go.'

He steps away, invites me to look.

'What am I looking at?'

'That diffuse cloud in the middle of your view. The Andromeda galaxy. It's the Milky Way's nearest neighbour. About 2.5 million light years from earth.'

And then it hits me. A kind of fatigue, born of over-exposure to bewildering numbers. Umpty-thrillion light years to this place; thrumpty-gazillion light years to that other place. If you laid all the football pitches and all the buses in Wales and Belgium next to Nelson's column, you'd need zumpty-megadillion of them to walk to the moon.

I thank him, and wander away, tilting my head back and scanning the dark sky vaguely. I have had a good time. I have learned loads of stuff, and some of it even stayed in my head. And I did, against the odds and the weather forecasts, do some stargazing.

I sit on a bench and allow the night sky to infiltrate my being. After a while they gather us together and bring the evening to a close, and I drive home, trying to concentrate on the road and not the infinite sky above.

Another place, another telescope. A different kind. No lenses, no mirrors. You don't squint through this one.

Jodrell Bank, in Cheshire. Set in thirty-five acres of entirely pleasing gardens and woods, and looking at the farthest reaches of the observable universe since 1945.

We walk round the Lovell Telescope, marvelling at the sheer beauty of the structure, the intricacy of its latticed metalwork, the impression it gives of being some sort of gut-churning fairground ride or an impossibly expensive scaled-up executive desktop toy, somehow both monumental and delicate at the same time.

It is a thing of great beauty. Had it been built purely for decoration I would still walk in its shadow with admiration – factor in its ability to register an alien fart in the Stingray Nebula and I'm solid gone.

Looking at stars through a reflecting telescope, seeing them with my own eyes, somehow gave me a thread of comprehension to cling to. I could see a tiny dot of light, so it must exist. And when the details were explained to me, I could at least pretend to understand while silently trying to picture a long series of noughts in my head and getting no further than 'a very very long way away'. But at least it felt real.

It's different with a radio telescope. This feels more like science fiction. I get that it's paraboloid because it's focusing the signal – it's a concept I became familiar with in my childhood when trying to record the song of a blackbird by nestling a cheap microphone in the cosy embrace of a dustbin lid. But how they decipher the radio signals when they have them is, as it should be, beyond me. And as with anything space-related, the question most likely to burst from my lips is an exasperated, 'But how do you *know*?'

And of course the answer always distils down to 'Science, and a lot of hard work by very clever people.'

I'm distracted – oh so easily distracted – by jackdaws. They nest under the telescope, the many ledges and crannies afforded by the supporting structure ideal for their purposes. I watch them tumbling and chasing each other, and hear their familiar *chack*ings, and I think, jackdaws I can deal with – they're on a scale I understand.

And then we look at the other things the site has to offer, particularly enjoying the delicious playfulness of standing by a parabolic reflector and hearing my son whisper into its twin thirty yards away, his voice appearing in my head as if he were standing next to me – a well-established acoustical trick, but great fun to try out anyway.

And when we're done with that, and feel like getting away from the milling throngs, we wander away from the telescope, through the grounds. I feel the warmth of the sun, and try hard not to think of it in terms of physics and maths, but merely to feel it as a pleasing sensation on my nape.

There is a bird hide in the woods. We sit for a while as blue tits and chaffinches come and go. There's the occasional rustle and flurry in the undergrowth, but I don't feel the need to dig out the binoculars and look for it. It isn't one of those days. It's a 'wandering through the woods with your family, allowing the time to pass slowly and uneventfully, occasionally admiring a butterfly, and simply enjoying the warmth and the light and the general ooh-ness of things while quietly turning over the vastness of the universe and the mystery of existence in your head, but not worrying too much about it' kind of day.

No doubt German has a word for it.

We stumble on a place. A clearing. A group of about twenty people, loosely gathered around what turns out to be a beekeeping demonstration. Information is imparted about the workings of a hive, the role of the queen and the workers and the drones and all sorts of fascinating stuff that I vaguely know already. I half focus on it, allowing the words to drift pleasantly into my head without feeling the need to retain the information. They quell incipient bee rebellion with a pleasant-smelling smoke, and gently warn us that sometimes bees might land on us. This is normal, they emphasise, and the main thing is not to panic, and above all not to flap, and certainly not to run, because that will cause the bee to emit a chemical summoning its colleagues to attack its attacker.

A bee lands on someone. They panic. They flap. They run. And while they do that, I think uncharitable thoughts about human intelligence and people's inability to take the advice of experts. To calm myself down I take a slow walk away from the action and into a nearby clearing where there is nothing but a chiffchaff singing its repetitive song to accompany my thoughts.

This feels like an appropriate place to end my journey, started back on the mean streets of West Norwood and taking me from the kitchen sink to the cosmic void. There's been wildlife galore – things I might never have dreamed of looking at if I'd stuck to my primary interest in all things avian. The spider in the sink, reminding me of the importance of not overlooking the apparently mundane; the Perfectly Normal Tree, harbouring invisible miracles; the lizard on the boardwalk, throwing me back to prehistory; the gorilla in the zoo, making me examine the ethics of our relationship with wild things; the beluga in the Thames, an alien life form visiting from far-off lands.

The places I've visited – working my way from the urban and domestic, through the rarefied atmosphere of museums, out through zoos and nature reserves to the wild and untamed – have yielded their treasures, whether in the dubious form of a sharp peck to the head from an Arctic tern or the more passive pleasures of a frosty, misty, jaunty-whistle-inducing morning above Selborne. And the people historically associated with those places – the great and the good – have been an inspiration. They all have something to offer – Darwin's patience, White's perception, Bewick's eye, Lemon's doggedness – and from them I've learned to slow down, to delve deeper, to seek out the unusual among the commonplace, to look, look again, look better. And by looking, I learn to see, to appreciate, to understand just a little more.

Wherever I've gone, people have played their part, whether fully absorbed by nature or indifferent to it, from beginners to experts and every stop between. Sometimes, it has to be admitted, they've done this by the simple expedient of being absent, allowing me the selfishness of communing silently with the natural world. But sometimes their contribution has been invaluable: the boy in thrall to the dolphins on the Farne Islands boat; Tattoo Guy, with his love of pigeons and his non-biting dog; Mike, indulging me with the whitethroat; and yes, even Ms YooHoo There's the Eagle. Seeing their reactions, their relationships, their different ways of being in nature – from silent to noisy, mildly curious to thoroughly absorbed – has made me realise how widely people's relationships with nature vary. For some, it is healing, an essential part of their daily struggle with the difficulties of existence; for others, merely a conduit to joy; for many people it's an incidental thing, an occasional pleasure to be dipped into like a bedside anthology; and for yet more it is scary, alien, to be avoided at all costs.

But for all these differences, for all that there are as many relationships with nature as there are people, there is a kind of unity to it all. Because we are part of it, not separate from it. Weird as it might seem, all these magnificent creatures – from nematode worm to murderous jeeb – are our cousins, however distant. Darwin saw this. When he wrote about nature, he included us. We all came from the same ancestor, and we've all survived this far. We are all in this together. And the more we understand our family, the closer we get to it, the more likely we are to take care of it.

A movement catches my eye. A fluttering, scattery, apparently random movement, as of one not quite in control but going there anyway.

A common blue butterfly.

Through a gap in the trees I glimpse the telescope, quietly exploring the furthest reaches of the known universe. A stray bee buzzes around my head, then lopes off back to the hive, there to perform unfathomable miracles. And I think of it all, from bees to Betelgeuse – all made from the same stuff, repurposed stardust.

For a very long time we thought we were the centre of it. But enlightenment gradually dawned – more gradually to some than to others – and we know now, more or less, where we stand.

We are an impossibly tiny speck. And that's fine. And if it scares the bejeezus out of some people, I happen to find it somehow reassuring. Because all these encounters – the butterfly on the flower, the eagle over the water, the light from a distant star and all the rest of it – represent the slotting-in of a tiny piece of our universe jigsaw. We'll never come close to finishing it, but nonetheless we persevere, constructing our own little corner of it in the hope that one day we'll be able to stand far enough away for some of it to make sense.

Look. Look again. Look better.

ACKNOWLEDGEMENTS

This book is, as you'd hope, all my own work. But it would be nothing without the efforts of generations of naturalists on whose hard labour and wisdom I drew; not just the people featured in the book, but countless others, experts in their fields, often breaking new ground – their dedication remains a wonder and inspiration to me.

Of the people I encountered on my journey, the majority remained unwitting, anonymous accomplices. They are presented, more or less, as I saw them, and I offer them my apologies and thanks. Please bear in mind that I mean well.

Behind the scenes, May Webber shared her knowledge of butterflies and moths with generosity and enthusiasm; Iain Green remains my first port of call for photography expertise; Emma Mitchell is a source of inspiration and friendship, and, more specifically, pointed me towards Anna Atkins; Josie George quietly gets on with the business of being brilliant, but also inspires me to look closely at things; Nic Wilson kindly read the John Clare chapter and guided me in the right direction; Nick Hely-Hutchinson did the same with the short passage on disability; David Darrell-Lambert told me where to go on Skye; and Richard Brown and Giselle Eagle were the embodiment of generosity during my stay on Skokholm.

My sister-in-law Laura Pritchard read early drafts and, as always, made them better. And of course my wife Tessa not only gives me unflinching support but was invaluable when it came to all things gardeny. My son Oliver does the same, but without the gardening bit.

To the people at Elliott & Thompson – Simon, Jennie, Sarah, Pippa – a simple and heartfelt thank you for letting me do it.

INDEX